T0235432

Science, Freedom, Democracy

This book addresses the complex relationship between the values of liberal democracy and the values associated with scientific research. The chapters explore how these values mutually reinforce or conflict with one another, in both historical and contemporary contexts.

The contributors utilize various approaches to address this timely subject, including historical studies, philosophical analysis, and sociological case studies. The chapters cover a range of topics including academic freedom and autonomy, public control of science, the relationship between scientific pluralism and deliberative democracy, lay-expert relations in a democracy, and the threat of populism and autocracy to scientific inquiry. Taken together the essays demonstrate how democratic values and the epistemic and non-epistemic values associated with science are interconnected.

Science, Freedom, Democracy will be of interest to scholars and graduate students working in philosophy of science, history of philosophy, sociology of science, political philosophy, and epistemology.

Péter Hartl is a research fellow at the Institute of Philosophy, Research Centre for the Humanities, Budapest, MTA BTK Lendület Morals and Science Research Group. His research focuses on epistemology and the history of philosophy (Hume, Michael Polanyi). He has published papers on Polanyi, Hume, and modal epistemology. He co-edited "The Value of Truth" special issue for *Synthese*. His monograph on Hume is under contract.

Adam Tamas Tuboly is a postdoctoral researcher at the Institute of Philosophy, Research Centre for the Humanities, Budapest, MTA BTK Lendület Morals and Science Research Group, and a research fellow at the Institute of Transdisciplinary Discoveries, Medical School, University of Pécs. He works on the history of logical empiricism and has edited numerous volumes on it.

Routledge Studies in the Philosophy of Science

For more information about this series, please visit: https://www.routledge.com/Routledge-Studies-in-the-Philosophy-of-Science/book-series/POS

Science, Freedom, Democracy

Edited by
Péter Hartl and Adam Tamas Tuboly

Routledge
Taylor & Francis Group

NEW YORK AND LONDON

First published 2021
by Routledge
52 Vanderbilt Avenue, New York, NY 10017

and by Routledge
2 Park Square, Milton Park, Abingdon, Oxon OX14 4RN

Routledge is an imprint of the Taylor & Francis Group, an informa business

© 2021 Taylor & Francis

The right of Péter Hartl and Adam Tamas Tuboly to be identified as the authors of the editorial material, and of the authors for their individual chapters, has been asserted in accordance with sections 77 and 78 of the Copyright, Designs and Patents Act 1988.

Library of Congress Cataloging-in-Publication Data
Names: Hartl, Péter (Péter Zoltan), editor. | Tuboly, Adam Tamas, editor.
Title: Science, freedom, democracy / edited by Péter Hartl and Adam Tamas Tuboly.
Description: New York, NY : Routledge, 2021. |
Includes bibliographical references and index.
Identifiers: LCCN 2020045020 (print) | LCCN 2020045021 (ebook) |
ISBN 9780367418175 (hardback) | ISBN 9780367823436 (ebook)
Subjects: LCSH: Science--Social aspects. | Science--Political aspects. |
Research--Social aspects. | Research--Political aspects. |
Academic freedom. | Democracy and science. | Pluralism.
Classification: LCC Q175.5 .S29842 2021 (print) |
LCC Q175.5 (ebook) | DDC 303.48/3--dc23
LC record available at https://lccn.loc.gov/2020045020
LC ebook record available at https://lccn.loc.gov/2020045021

ISBN: 978-0-367-41817-5 (hbk)
ISBN: 978-0-367-70400-1 (pbk)
ISBN: 978-0-367-82343-6 (ebk)

Typeset in Sabon
by Taylor & Francis Books

To my mother (P.H.)
To Gabó, Sára, Róza (A.T.T.)

Contents

viii *Contents*

1 Introduction

Péter Hartl

RESEARCH CENTRE FOR THE HUMANITIES, INSTITUTE OF PHILOSOPHY, MTA BTK
LENDÜLET MORALS AND SCIENCE RESEARCH GROUP

Adam Tamas Tuboly

RESEARCH CENTRE FOR THE HUMANITIES, INSTITUTE OF PHILOSOPHY, MTA BTK
LENDÜLET MORALS AND SCIENCE RESEARCH GROUP/ INSTITUTE OF TRANSDISCIP-
LINARY DISCOVERIES, MEDICAL SCHOOL, UNIVERSITY OF PÉCS

The concepts of "science", "freedom", and "democracy" are often recognized as deeply connected with one another, even though each is subject to various interpretations and involves many complications. A systematic account of these topics should clarify how the values of liberal democracy and the values associated with scientific research can either mutually reinforce or conflict with one another. By integrating historical, analytical, and empirical approaches, this book aims to provide a comprehensive philosophical overview of these complex questions from different perspectives. To this end, the term "science" is considered as broadly as possible, and also covers the social sciences and humanities.[1]

This volume consists of essays by internationally renowned and emerging young scholars that investigate interrelated questions, including the nature and limits of academic freedom, the social responsibility of scientists, the moral and social values of the scientific community in a democracy, methodological and value pluralism in science policies, the relation between participatory democracy and science, and the threats of populism and autocracy. The contributions also cover the topics of philosophical criticism of totalitarian control of science, various philosophical theories of democracy, public reason, the politicization of science, and epistemic injustice. Some contributions to the volume rely on classic approaches from the history of philosophy (for example those of Max Weber, Robert Merton, and Michael Polanyi) to examine present-day challenges to academic freedom and democracy. Others utilize empirical case studies in their philosophical argumentation about the role of science in democratic and free societies.

More specifically, the chapters in this volume focus on three main areas. Part I ("Academic Freedom and Other Values in Science and Society") focuses on historical and contemporary analyses of different conceptions of academic freedom and its relation to other social and moral values of the scientific community and liberal democracy. Academic freedom can be defined as the freedom of either individual researchers or scientific communities to choose their own topics, problems, and methods (Polanyi 1951, 33).

Another aspect of academic freedom is autonomy: scientific institutions and researchers should be largely independent of external, non-epistemic influences (whether financial, political, or ideological). Especially since the Second World War, academic freedom has been perceived as virtually boundless, with external, non-scientific aims regarded as a threat.[2] In this view, academic freedom should be constrained only by basic ethical considerations (Bridgman 1947). Accordingly, the social and moral values associated with academic freedom are the fundamental values of free society, and arguably of any democracy. Democracy and free society are modeled after a more or less idealized picture of the scientific community, typically referred to as "the republic of science" (Polanyi 1962). Such uncompromising defense of academic freedom is proposed as an alternative to the Nazi and Stalinist totalitarian control of science. Moreover, following Francis Bacon's optimistic ideas, it has been a popularly held view in the post-war period that virtually unlimited academic freedom is – in most cases, if not always – useful for society, based on the belief that scientific research freed from constraints will inevitably produce useful knowledge and beneficial technology (Bush 1945).

However, such optimism about scientific freedom and social prosperity has since been questioned. Some have also criticized the classic ideal of academic freedom and the optimism attached to its value, arguing that to some extent, external, even political, control of science is necessary, and that almost unlimited freedom without public accountability is not desirable in modern democracies (Edsall 1975; Douglas 2003; Kitcher 2011). As such, public accountability of scientists is necessary, and the harmful social consequences of technological progress often associated with boundless freedom in economics and science need to be critically examined. Nevertheless, it remains an open question how to draw normative distinctions between necessary or acceptable and unnecessary or inadmissible forms of public control of scientific research.

The contributors to this part of the volume (Phil Mullins, Péter Hartl, Heather Douglas, and Janet Kourany) focus on the following problems: the nature and the limits of academic freedom, in what sense academic freedom is necessary for a free society, to what extent academic freedom can or should be limited, and which other values beside academic freedom should guide scientific communities and individual scientists. Whereas Mullins and Hartl put greater emphasis on the nature and the value of academic freedom in liberal democracies, Douglas and Kourany examine the role of other values (such as social prosperity, equality, accountability, and responsibility) for scientific communities and society as a whole.

Phil Mullins's essay ("Michael Polanyi's Post-Critical Philosophical Vision of Science and Society") reconstructs Michael Polanyi's critical and constructive philosophy, with a focus on his social and political ideas and how they relate to his account of science. Mullins examines why Polanyi thought that modern philosophy ultimately led to a false conception of science and gave rise to nihilism, violence, and totalitarianism in the twentieth century.

The chapter presents Polanyi's constructive "post-critical" alternative perspective focused on a discovery-centered account of science and a broadly reframed understanding of human knowledge. Mullins shows how Polanyi interweaves his account of science with ideas about how the social and political organization of society could best promote science according to the vision of a "society of explorers".

Péter Hartl's chapter ("The Ethos of Science and Central Planning: Merton and Michael Polanyi on the Autonomy of Science") discusses in historical context how Merton's and Polanyi's conceptions of values, norms, and traditions in science (i.e., "the ethos of science") laid the ground for their defense of academic freedom and a free society. Hartl argues that Merton's and Polanyi's defense of freedom of science and free society against totalitarianism have much in common, even though Polanyi himself was critical of Merton's sociology. Polanyi's criticism of Merton, as Hartl shows, arises from a misreading of Merton's position. Both thinkers argue that freedom of science and a free society depend on each other, and both share the idea that the ethos of science exemplifies the values of a free society as a whole. They both believe that centralized control is motivated by the illiberal idea that the state is solely responsible for the welfare of society. For Polanyi, scientific inquiry is a spontaneous order whose development would be inevitably paralyzed by a central and hierarchic organization. Hartl highlights problems with both Merton's and Polanyi's views and suggests some modifications thereof, but concludes that Merton's and Polanyi's warnings offer valuable lessons for the present day, as the basic motives underlying the attempts by governments to control science are still with us in the context of populist and authoritarian politics.

Heather Douglas's chapter ("Scientific Freedom and Social Responsibility") analyzes scientists' awareness of their responsibilities toward the societies in which they pursue their research, which has recently increased. The chapter examines the shift from the view of the social responsibility of scientists that predominated in the second half of the twentieth century and investigates how responsibility is now yoked to the freedom to pursue scientific research, rather than opposed to such freedom. First, Douglas describes this change and its causes; second, she addresses the fact that much of our institutional research oversight infrastructure was put in place with a now-outdated understanding of the societal responsibilities and freedoms of scientists; and finally, she makes some recommendations on how to reform the structures we have in place to be more in tune with the current conceptualization of scientific freedom and social responsibility.

Janet Kourany's chapter ("Bacon's Promise") critically examines the modern view (traced back to Bacon) that scientific discoveries will inevitably produce prosperity for everyone. The chapter criticizes Vannevar Bush's optimistic "Baconian" concept, according to which science, if it is financially supported and left free of any social control, will produce social progress in every respect (health, happiness, security, and prosperity). Although Kourany does not deny that science has produced significant benefits for modern society, she argues that the consequences of scientific and technological progress are rather mixed. She relies on case studies from the food industry and medical research to show that certain scientific-technological achievements have caused serious health problems. She also analyzes how technology and science-driven economic growth have contributed to increasing inequality in the distribution of wealth. Kourany argues that instead of taking for granted that academic freedom without external control has beneficial consequences only, we should reconsider how to supervise and regulate (often very costly) scientific and technological projects and how to hold researchers publicly accountable.

Part II ("Democracy and Citizen Participation in Science") focuses on various definitions of "democracy" and the role that certain democratic values, notably participation and deliberation, could and should play in science policy and how and to what extent science should be "democratized".[3] The scientific community can serve as an inspiration for certain ideals of democracy, since the scientific ethos is often seen as essentially liberal, anti-authoritarian, critical of ideologies, and opposed to any discrimination based on race, gender, class, or otherwise (Sarton 1924; Merton 1973; Popper 1945). On the other hand, some scholars have contrasted these somewhat idealized epistemic and social norms of scientific communities with partisan politics and have argued in favor of rule by experts, as opposed to democratic elections based on universal suffrage (Comte 1875–1877; Brennan 2016). Yet others have criticized hierarchies and conservatism in science and have called for egalitarianism both in scientific institutions and in society (Fuller 2000; Jarvie 2001).

At any rate, the term "democracy" notoriously has many meanings and is often used in conjunction with different quantifiers (Shapiro and Hacker-Cordón 1999; Cunningham 2002). Participatory democracy is frequently contrasted with representative democracy, and since Rousseau there has been a long tradition of considering direct, participatory democracy, where citizens themselves take part in deliberative processes, as the genuine, or at least the more authentic form of democracy, rather than rule by a political class (Arendt 1998, 31–33, 220–226; 1990, 30–33). The core concept of deliberative democracy, developed mainly by Rawls and Habermas, is "public reason", which encapsulates moral or political norms that can be justifiably imposed upon free and equal individuals (Chambers 1996; Gauss 2010, 184; Habermas 1998, 41–44; Habermas 1996; Rawls 1996, 36–37, 55–58). How these norms should be determined, to whom these rules are justifiable, and whether deliberation requires equal consideration of all perspectives or should assign greater weight to expert opinions is still the subject of much debate (Bohman 1996; Cohen 2008; Enoch 2015). Complete deference to scientific expert opinions seems to conflict with at least some conceptions of public reason or, at worst, with democratic principles in general (Kitcher 2011, 20–25; Mulligan 2015).[4]

Although the statement that "scientific truth is not a matter of the decision of the majority" is hardly debatable, many scholars have argued that scientific research should be supervised, if not to some extent controlled, either by democratic institutions or directly by the public in a well-functioning democratic society (Brown 2009). Arguably, the most elaborate analysis of the role of science in a democratic society is proposed by Kitcher (2001; 2011) who believes that scientific research and institutions must serve the goals of democracy – in other words, that public knowledge should be shaped to promote democratic and egalitarian norms and values. The question arises to what extent these scientific institutions and decisions on scientific research agendas can be properly called democratic. Moreover, it needs to be examined to what extent the general public, juries of citizens, or elected politicians could and should decide on scientific research agendas, scholarships, goals of scientific research, and moral-social questions related to scientific knowledge (Brown 2013; Leschner 2003; Est 2011; Ruphy 2019).

The chapters of Part II, by Hans Radder, Hugh Lacey, and Dustin Olson, contribute novel insights to the debates around the following questions: which approaches to science should be fostered in the context of upholding democratic values (in particular, the values of participatory and deliberative democracy); to what extent is citizen participation in decisions concerning science policies and scientific research legitimate or useful; and how should democratic institutions support public decision making by mitigating the distortions of scientific expert opinions.

Hans Radder's contribution ("Which Science, Which Democracy, and Which Freedom?") has two aims. First, to elaborate on the notions of science, democracy, and freedom in the context of the diversity of the sciences and their relations to technology. Radder's account of democracy combines the features of voting, deliberation, and the separation of the executive, legislative, and judiciary powers of the state, as well as appropriately inclusive elections. His take on the notion of (negative and positive) freedom acknowledges its significance as a general regulative value, while also emphasizing its entanglement with sociopolitical interpretations and contextualizations. The second aim of the chapter is to discuss a range of examples of the actual and desirable relations between science, democracy, and freedom. Radder's examples include the significance of the double hermeneutic for the human sciences; an analysis and assessment of the development of the Dutch National Research Agenda as a form of citizen participation in science; a discussion of the principal legal and social aspects of academic freedom; and, finally, a defense of the thesis that it is in the public's interest to guarantee the freedom to pursue basic academic research.

Hugh Lacey's chapter ("Participatory Democracy and Multi-Strategic Research") examines what approaches to science should be fostered in the context of upholding democratic values. The chapter analyses two conceptions of democracy: "representative democracy" and "participatory democracy", and two of scientific research: "decontextualizing research" (i.e., research into which decontextualizing strategies are adopted); and "multi-strategic research" (i.e., research that allows for a pluralism of methodological strategies). Lacey discusses connections between representative democracy and decontextualized strategies in research. Moreover, he examines the possible consequences for science of recent threats to democratic values and institutions in a number of countries. Lacey's thesis is that upholding the values of participatory democracy fosters multi-strategic research, which in turn buttresses the traditional ideals of modern science.

Dustin Olson's contribution ("Public Opinion, Democratic Legitimacy, and Epistemic Compromise") relies on a recent example from US politics – the climate change debate in the 2010 mid-term elections – to show how public opinion is shaped through the exploitation of epistemic interdependence and partisan bias. Prior to the election, public opinion on climate change was subject to several willfully disseminated distorting influences, which had a significant impact on the outcome and compromised its democratic legitimacy. To defend democratic legitimacy, Olson demonstrates how we can re-evaluate and update the parameters under which the liberal institutions necessary for informed public opinion operate in light of our current political and epistemic needs. Moreover, Olson argues that democratic institutions have moral, political, and epistemic obligations to facilitate public reason and to mitigate the willful dissemination of distorting influences on democratic decision making.

Part III ("Freedom and Pluralism in Scientific Methodology and Values") focuses in general on how pluralism about values in certain domains should influence our understanding of scientific research and guide science policy in a free society. According to the classic idea of value-free science, scientific opinion must not be influenced by value judgments other than epistemic values (Nagel 1961). Hume and the positivists argued that value judgments do not express propositions that can be true or false, and that values are therefore not subject to scientific investigation (Hume 1975, 172–173; Hempel 1965, 85–86).[5] Since the 1960s and 1970s, the defenders of the "strong" program of sociology of science, as well as various social constructivists (Barnes, Bloor, and Henry 1996; Latour and Woolgar 1979), have argued for the opposite view: given that external, non-epistemic values (such as political, religious, or ideological ones) inevitably determine scientific theory choices, the value-free ideal has to be rejected and a relativistic understanding of scientific knowledge should be adopted. However, in recent years, both the value-free, objectivist idea of science and relativistic social constructivism have been scrutinized and reconsidered, if not rejected. In their place, some scholars have proposed a position between the two extremes that recognizes and approves of non-epistemic values in science but avoids relativism (Lacey 1999; Douglas 2009). Others have rejected the assumption that epistemic values should be taken as a priori and have adopted a pragmatist stance about scientific and democratic values (Brown 2017).

If the importance of both epistemic and non-epistemic values in science is recognized, it is also clear that there are many such values, including methodological, moral, practical, and political ones. In all domains, it can be plausibly argued that there is no single value that could or should influence or guide scientific institutions and science policies, and that a pluralist approach should be adopted instead. It also has to be kept in mind that pluralism can have multiple contexts; given that many authors reject the very strict and sharp distinction between epistemic/non-epistemic values (or deny its significance), pluralism might also answer their concerns by calling attention to the idea that values (whatever their category) always come in the plural. Furthermore, pluralism in science and values would reflect, in a more structured way, the actual practice of many scientific fields. It thus provides a strong theoretical device for engaging with actual problems and challenges, while still leaving open the possibility for normative issues to be resolved by local reductions to more monolithic value discourses (for further references, see Elliott and Steel 2017).

The contributions in Part III, by Jeroen Van Bouwel and Lidia Godek, investigate the following questions: which types of values are legitimate in guiding scientific communities and decisions about science policy in a free society, and on which domains and to what extent can value pluralism be used in science policy making alongside our theoretical understanding of science.

Jeroen Van Bouwel ("Are Transparency and Representativeness of Values Hampering Scientific Pluralism?") examines under what conditions the influence of values on science is justifiable. The chapter analyzes two of three conditions proposed by Elliott: that influences should (1) be made *transparent* and (2) be *representative* of our major social and ethical priorities. By analyzing transparency initiatives in political science, Van Bouwel argues that the first condition has benefits but also drawbacks, in particular in relation to initiatives for fostering scientific pluralism. Moreover, he maintains that the second condition might help us to answer which or whose values scientists should legitimately follow. One way to understand the second condition is that the values used in science should be representative and democratically legitimate. To further elaborate on Elliot's views, Van Bouwel relies on economic models and proposes an alternative account of representativeness, rejecting conceptions that hamper scientific pluralism. Instead, Van Bouwel contends that a reflection on social and epistemic practices in science shows that commitments to transparency and representativeness should include a commitment to agonistic, pluralist democracy and scientific-methodological pluralism. The chapter illustrates how questions about scientific plurality and values are interrelated and why a commitment to pluralism is beneficial for both democracy and science as an epistemic enterprise.

Lidia Godek's chapter ("Max Weber's Value Judgment and the Problem of Science Policy Making") argues that to understand Weber's concept of value judgments and its practical implications, we must analyze his meta-philosophical thesis of "reference to values". Godek begins by discussing the question of how we should understand the role of value judgments and to what extent they are acceptable in scientific practice and science policy making. The chapter distinguishes two levels on which Weber's conception of value judgments can be analyzed: (a) *the methodological level* relating to the choice of the subject matter of scientific inquiry (a reference to values, value axioms, and ideal types); and (b) *the worldview level*, including the postulate of value freedom as an assumption of modern science. For Weber, value judgments cannot be reduced to the practical value judgments of scientists. Godek reveals how, in Weber's theory, such a plural perspective on value judgments is manifested in the value of the vocation of scientists and the values associated with the institutionalization of science. Finally, Godek considers the consequences of the institutionalization of values and develops three different models of science policy making, namely a *regulative*, a *protective*, and an *integrating* approach.

Acknowledgements

The essays in this volume (with two exceptions) were originally presented at the international conference on *Science, Freedom, Democracy* on July 8–9, 2019, at the Hungarian Academy of Sciences, Institute of Philosophy, Budapest, Hungary. The conference and the publication of this book were supported by MTA BTK Lendület Morals and Science Research Project, Hungarian Academy of Sciences – International Conference Fund and the Research Centre for the Humanities. The volume contributes to the research program of the MTA BTK Lendület Morals and Science Research Project (Hartl-Tuboly) and to the MTA Premium Postdoctoral Fellowship program (Tuboly). We are indebted to our authors for their participation and the stimulating discussions both at the conference and in this volume. We are also grateful to our reviewers and editors at Routledge for their kind work and help, especially during the recent global health crisis.

Notes

1 As is the case with this volume, it is often sociologists of science and philosophers of science who take up the topics of academic freedom and public engagement – despite the fact that these also deserve the attention of political and social philosophers. Although the perspectives of political and social philosophers could enrich these discussions, our aim here is only to lay out a tapestry of recent issues and to foster discussions among philosophers that bear straightforward relevance to contemporary discourses.

2 After the Second World War, especially during the McCarthy era, the Cold War transformed many of the sciences and their public relations to society. Philosophy of science, largely imported from socialist Red Vienna, quickly lost its sociopolitically engaged character and fell back on more neutral and technical subtleties. See Reisch (2005).

3 It should be noted here that topics such as "public engagement", "citizen participation", and similar problems of the relation between science and democracy are

relatively recent issues in the philosophy of science. While it is generally acknowledged that philosophy of science was a socially engaged and politically active field before the Second World War (Howard 2003), it was mainly restricted to the epistemic and moral accountability of scientists and was motivated by a certain Enlightenment vision of the public's empowerment. Current debates about values in science and society are more focused, specialized, and pay greater attention to actual regulations and norms.

4 In this respect, one might suspect a broader narrative where cultural factors play a key role. During the early decades of the twentieth century, leading scientists often produced detailed works about the philosophical implications of their scientific achievements (see Arthur Eddington, James Jeans, J.B.S. Haldane, J.D. Bernal, and others). These works achieved best-seller status and were also widely influential among non-scientists. But their impact (as documents of the cultural sensitivity of their era) extended rather to the level of concepts and worldviews and not to policies, public governance, or the shaping of political norms. One might thus suspect that the notion of "experts" had a different meaning then and now, and that the questions of philosophy of science evolved and developed along different lines.

5 It should be noted, however, that while many logical positivists accepted the idea of non-cognitivism (namely the view that value judgments do not express genuine descriptive propositions and are thus not truth-apt), they still considered moral and political discourses to be of utmost importance and hence tried to find alternative ways to engage with such issues; some of them delivered popular talks about the everyday (and thus moral) relevance of science in adult education centers; other organized exhibitions, museums, and films; and a third group transferred the subject to psychology or mathematics (as in game theory). Thus, against general wisdom, it is important to be aware of how philosophers (of science) let ethical and socio-political issues slip away from the canon for decades.

References

Arendt, H. 1990. *On Revolution*. London: Penguin Books.

Arendt, H. 1998. *The Human Condition*, 2nd edition. Introduction by Margaret Canovan. Chicago and London: University of Chicago Press.

Barnes, B., D. Bloor, and J. Henry. 1996. *Scientific Knowledge: A Sociological Analysis*. Chicago: University of Chicago Press.

Bohman, J. 1996. *Public Deliberation: Pluralism, Complexity, and Democracy*. Cambridge, MA: MIT Press.

Brennan, J. 2016. *Against Democracy*. Princeton, NJ: Princeton University Press.

Bridgman, P.W. 1947. "Scientists and Social Responsibility." *The Scientific Monthly* 65 (2): 148–154.

Brown, Mark B. 2009. *Science in Democracy: Expertise, Institutions, and Representation*. Cambridge, MA: MIT Press.

Brown, Matthew J. 2013. "The Democratic Control of the Scientific Control of Politics." In *EPSA11 Perspectives and Foundational Problems in Philosophy of Science*, edited by V. Karakostas and D. Dieks, pp. 479–491. Dordrecht: Springer.

Brown, Matthew J. 2017. "Values in Science: Against Epistemic Priority." In *Current Controversies in Values in Science*, edited by K.C. Elliott and D. Steel, pp. 64–78. New York and London: Routledge.

Bush, V. 1945. *Science: The Endless Frontier*. Washington, DC: US Government Printing Office.

Chambers, S. 1996. *Reasonable Democracy: Jürgen Habermas and the Politics of Discourse*. Ithaca, NY: Cornell University Press.

Cohen, J. 2008. "Truth and Public Reason." *Philosophy & Public Affairs* 37 (1): 2–42.

Comte, A. 1875–1877. *System of Positive Polity*. London: Longmans, Green.

Cunningham, F. 2002. *Theories of Democracy. A Critical Introduction*. New York and London: Routledge.

Douglas, H. 2003. "The Moral Responsibilities of Scientists: Tensions between Autonomy and Responsibility." *American Philosophical Quarterly* 40 (1): 59–68.

Douglas, H. 2009. *Science, Policy, and the Value-Free Ideal*. Pittsburgh, PA: Pittsburgh University Press.

Edsall, J.T. 1975. *Scientific Freedom and Responsibility*. Washington, DC: AAAS.

Elliott, K.C., and D. Steel (eds.). 2017. *Current Controversies in Values and Science*. New York and London: Routledge.

Enoch, D. 2015. "Against Public Reason." In *Oxford Studies in Political Philosophy*, vol. 1, edited by D. Sobel, P. Vallentyne, and S. Wall, pp. 112–142. Oxford: Oxford University Press.

Est, R. 2011. "The Broad Challenge of Public Engagement in Science." *Science and Engineering Ethics* 17 (4): 639–648.

Fuller, S. 2000. *The Governance of Science: Ideology and the Future of the Open Society*. Buckingham: The Open University Press.

Gauss, G. 2010. *The Order of Public Reason: A Theory of Freedom and Morality in a Diverse and Bounded World*. Cambridge: Cambridge University Press.

Habermas, J. 1996. *Between Facts and Norms: Contributions to a Discourse Theory of Law and Democracy*, translated by W. Rehg. Cambridge, MA: MIT Press.

Habermas, J. 1998. *The Inclusion of the Other: Studies in Political Theory*, edited by C.P. Cronin and P. De Greiff. Cambridge, MA: MIT Press.

Hempel, C.G. 1965. "Science and Human Values." In *Aspects of Scientific Explanation and other Essays in the Philosophy of Science*, pp. 81–96. New York: Free Press.

Howard, D. 2003. "Two Left Turns Make a Right: On the Curious Political Career of North American Philosophy of Science at Mid-century." In *Logical Empiricism in North America*, edited by A. Richardson and G. Hardcastle, pp. 25–93. Minneapolis: University of Minnesota Press.

Hume, D. 1975. "An Enquiry Concerning the Principles of Morals." In *Enquiries Concerning Human Understanding and Concerning the Principles of Morals*, edited by L.A. Selby-Bigge, 3rd revised edition by P.H. Nidditch, pp. 169–323. Oxford: Clarendon Press.

Jarvie, I.C. 2001. "Science in a Democratic Republic." *Philosophy of Science* 68 (4): 545–564.

Kitcher, P. 2001. *Science, Truth, and Democracy*. Oxford: Oxford University Press.

Kitcher, P. 2011. *Science in a Democratic Society*. New York: Prometheus Books.

Lacey, H. 1999. *Is Science Value Free? Values and Scientific Understanding*. London and New York: Routledge.

Latour, B., and S. Woolgar. 1979. *Laboratory Life: The Social Construction of Scientific Facts*, introduction by Jonas Salk. Beverly Hills, CA: Sage Publications.

Leschner, A.I. 2003. "Public Engagement with Science. (editorial)." *Science* 299 (5609): 977.

Merton, R.K. 1973. "Science and the Social Order." In *The Sociology of Science. Theoretical and Empirical Investigations*, edited and with an introduction by Norman W. Storer, pp. 254–267. Chicago: University of Chicago Press.

Mulligan, T. 2015. "On the Compatibility of Epistocracy and Public Reason." *Social Theory and Practice* 41 (3): 458–476.

Nagel, E. 1961. *The Structure of Science: Problems in the Logic of Scientific Explanation*. New York: Harcourt, Brace and World.

Polanyi, M. 1951. "Foundations of Academic Freedom." In *The Logic of Liberty. Reflections and Rejoinders*, pp. 32–49. London: Routledge and Kegan Paul.

Polanyi, M. 1962. "The Republic of Science: Its Political and Economic Theory." *Minerva* 1: 54–73.

Popper, K. 1945. *The Open Society and Its Enemies*, vols. I–II, 1st edition. London: Routledge and Kegan Paul.

Rawls, J. 1996. *Political Liberalism*. New York: Columbia University Press.

Reisch, G. 2005. *How the Cold War Transformed Philosophy of Science: To the Icy Slopes of Logic*. New York: Cambridge University Press.

Ruphy, S. 2019. "Public Participation in the Setting of Research and Innovation Agenda: Virtues and Challenges from a Philosophical Perspective." In *Innovation beyond Technology: Science for Society and Interdisciplinary Approaches*, edited by S. Lechevalier, pp. 243–263. Singapore: Springer.

Sarton, G. 1924. "The New Humanism." *Isis* 6 (1): 9–42.

Shapiro, I., and C. Hacker-Cordón (eds.). 1999. *Democracy's Value*. Cambridge: Cambridge University Press.

Part I

Academic Freedom and Other Values in Science and Society

2 Michael Polanyi's Post-Critical Philosophical Vision of Science and Society

Phil Mullins

MISSOURI WESTERN STATE UNIVERSITY

2.1 Introduction

Michael Polanyi is often called a "philosopher of science" or an "epistemologist" and these shorthand tags, used especially by philosophers, seem generally on target. However, what Polanyi called his "post-critical philosophy" is much richer than these tags suggest.[1] In this historically oriented discussion of Polanyi's thought, I show that there is much in Polanyi about social, political, and cultural matters and their relation to science and modern society. Polanyi's interdisciplinary writing articulated, on one hand, a critical account of the development of modernity, leading to what he called the "perilous state of the modern mind" (Polanyi 1965, 13). This critical account included his carefully developed perspective outlining how science and human knowing was misrepresented in the development of modern ideas, and how this misrepresentation has led to nihilism, violence, and totalitarianism in the twentieth century. But Polanyi also outlined a constructive alternative to this narrative. His "post-critical" perspective recasts much modern thought about scientific knowledge and knowing in general.[2] In Polanyi's thought, the critical and the constructive elements are woven tightly together, and this makes Michael Polanyi a difficult thinker, one whose significance has often been underestimated. Following a brief review of Michael Polanyi's life and work, I discuss a selection of provocative Polanyi comments about science and modern society that together outline the contours of both Polanyi's criticisms of modernity and what Polanyi called his "post-critical philosophy".

2.2 Crossing Disciplinary Boundaries: A Brief Overview of Polanyi's Life and Thought

In 1891, Michael Polanyi was born the fifth of six children in a prosperous, secular Jewish family in Budapest.[3] He received a superb early education in the humanities as well as in science. His family's prosperity declined after his father's bankruptcy in 1900 and death in 1905. Polanyi began the study of medicine at the University of Budapest in 1908. While he was still training to be a physician, he became deeply interested in physical chemistry and, in the summer of 1912 and then again in parts of 1913 and 1914, Polanyi studied physical chemistry at the Technische Hochschule in Karlsruhe, Germany. One of his early chemistry research papers was sent by a professor to Einstein, who approved it, and it was quickly published. As Polanyi (1975, 1151) later humorously described this early experience: "Bang, I was created a scientist". Polanyi finished his medical degree in April 1913, but he continued his research in chemistry while he practiced medicine in the Austro-Hungarian army during World War I. His health was delicate, and he was hospitalized for a period during the war and he had light duty at other times, and thus was able to pursue chemistry research projects, and to publish a few papers. One of these he eventually turned into a PhD thesis in chemistry and his degree was granted in July 1919. After the war, Polanyi was very briefly an official in the Hungarian Ministry of Health, but after the fall of the Liberal government, he resigned and left Hungary in late 1919 and returned to Karlsruhe and his work in physical chemistry.

Polanyi's impressive early research earned for him a position at the Kaiser Wilhelm Institutes near Berlin, which were among the world's best scientific research facilities in this period, and Polanyi there became a bright, rising star. In his 13 years in Germany, Polanyi produced outstanding scientific research and interacted with some of Europe's best scientists. In many ways, Polanyi's later philosophical ideas about the scientific enterprise were formed in these years (Nye 2011, 85–112). But even as a research scientist Polanyi was somewhat atypical since he worked in several different areas of chemistry.

In 1933, with the rise of the Third Reich, Polanyi took a position at the University of Manchester where he continued doing chemistry research for 15 years. In 1948, he exchanged his chair in chemistry for one in social studies in order to work in economics and philosophy and prepare for his upcoming Gifford Lectures. Even before he left Berlin, and certainly after his move to Manchester, Polanyi was already focusing on interests outside the natural sciences. He was deeply troubled by the changed world that World War I brought, and his writing in the thirties and forties reflects that Polanyi was struggling to understand where modern culture and politics, deeply influenced by science, had gone wrong. The first steps toward a new career outside chemistry came in the thirties and early forties when he produced an economics education film (Bíró 2020, 36–104) as well as notes, lectures, essays, and books that treated politics, economics, and particularly questions about "planned" science.

Especially after Polanyi exchanged his chemistry chair for a chair in social studies, he was very active as a public intellectual, scholar, and lecturer in the UK and the US. He later described his career in philosophy as something of an "afterthought" (Polanyi 1966, 3; see also Mullins 1997, 5–6 and Polanyi 1946/1964, 7–19, addition in 1964 reprint [hereafter SFS]), but by the middle of the last century Polanyi had turned from his earlier interests in economics and reforming liberal political ideas to probing broader philosophical questions about human knowing and scientific knowledge, the evolution of modern ideas, and the role of science in shaping modernity. Polanyi's broader inquiry grew out of his earlier interests, but also reflects that he came to see the problems of contemporary society as reaching beyond matters that economics and political discussions often probed.

However, Polanyi was never preoccupied with topics of current interest to most modern philosophers, and his writing was often ignored or attacked by them. He was not a guild member, but he articulated sharp criticisms of the modern philosophical tradition, focusing particularly on what he regarded as the obsession with objectivity; he attacked impersonal accounts of science that ignored the centrality of discovery. More generally, Polanyi criticized the narrowness of modern understandings of knowledge that overlooked the social and personal roots of knowing. Polanyi was steadfastly anti-Cartesian in his orientation and, as suggested above, wove with his sharp criticisms of modern ideas his alternative constructive "post-critical" account that reconceived the nature of knowledge and the activity of human knowing, showing the importance of tacit elements.

In 1951 and 1952, Polanyi gave his Gifford Lectures, two series of ten lectures, titled "Commitment, In Quest of a Post-Critical Philosophy", in which he begins to work out more fully his constructive philosophical perspective focusing on personal knowing. From these lectures, he produced, six years later, his magnum opus, *Personal Knowledge: Toward a Post-Critical Philosophy*. [4] In the sixties and early seventies, in many articles and several books, Polanyi refined his epistemological ideas and discussed problems of meaning in modernity. During the last 20 years of his life, Polanyi was involved in a staggering number of projects as a scholar and public intellectual.[5] He was in residence and gave public lectures at a number of European and North American universities. Over his career, Polanyi published about 200 technical scientific papers, but his diverse publications on economics, the history and philosophy of science, and society and culture are about as numerous.

2.3 Primary Social Values and Practices Required for the Flourishing of the Scientific Enterprise and a Modern Society Shaped by Science

In the troubled middle decades of the twentieth century, Polanyi unequivocally stated certain fundamental ideas about the social and cultural requirements that he believed are necessary if the scientific enterprise and modern culture shaped by the scientific enterprise are to flourish:

> Science, and generally the independent search for truth, is destroyed when political liberty falls. ... By its very nature ... [religious, political, or scientific] thought must claim superiority to temporal power ...
>
> [T]he link between science and liberty is completely reciprocal: while the profession of truth needs for its protection the free institution of democracy, these institutions themselves must decay and fall if people

abandon their belief in reason. The idea of liberty derives its strength from many roots but among these there is one most vital: the belief that men can reach a better understanding by free discussion, that in fact society can be continuously improved if public life is steadily guided by reasoned controversy.

(Polanyi 1937a, 710)

These carefully articulated comments are from a 1937 Polanyi letter published in *Nature*. This letter was written just after Polanyi returned from an international scientific congress in Paris, a meeting in which everyone present was mindful of the ways in which the work of those involved in the scientific enterprise hung in the balance as the Third Reich moved to dominate Europe.

Polanyi's letter makes clear several of his basic commitments. Science, and a society guided by science, must believe in truth and must be committed to an independent search for truth. Such an independent search requires political liberty and the institutions of democracy are best able to provide the liberty that nurtures and protects commitment to an independent search for truth. Finally, the independent search for truth will and should yield what Polanyi calls "reasoned controversy", that is, the vehicle through which knowledge grows and society takes steps forward. There is an ongoing public conversation in science and society – a reasoned and lively discussion – about truth, and that public conversation is grounded in political liberty.

2.4 Political Philosophy: Polanyi's Middle Position

In the thirties and forties in Britain, Polanyi regarded his own political philosophy as a middle position. His views countered, on one hand, what he called "extreme liberalism" (Polanyi 1940/1975, 56–58), which he sometimes characterized as a modern kind of barbarism, and, on the other hand, what he regarded as popular contemporary utopian ideas about centralization that he associated with both the Soviet Union and fascist governments. For Polanyi, ideas are important in society, and this is clearly the case in the domain of politics. Extreme liberalism (i.e., strict laissez faire economic liberalism) misunderstood unemployment and was superstitious in holding that "the market takes revenge on society for any interference with its mechanism". In Polanyi's account, extreme liberalism "was one of the most potent immediate reasons of the Nazi revolution, which might have been avoided by a policy of financial expansion" (Polanyi 1940/1975, 58). Polanyi found the "mystical element" in extreme liberalism akin to "obsessions of collectivists about the evil powers of the market" (Polanyi 1940/1975, 58–59).

He understood both communism and fascism to be forms of "totalitarianism", which as Polanyi defined it, was a peculiarly modern and currently appealing idea about governance in which the state is imagined as in principle completely responsible for the culture and welfare of its citizens, "a regime absorbing the whole life of the people, who live by it and live for it entirely" (Polanyi 1940/1975, 27). Under "totalitarianism", the public good is envisioned as achievable through the work of a comprehensively organized and directive bureaucracy. "Totalitarianism" asks why not deal with society as a whole and it proposes "comprehensive, provident action" and this, for modern people, "appeals also to moral feelings" (Polanyi 1940/1975, 28).

Polanyi rejected both "extreme liberalism" and "totalitarianism" and favored what he regarded as a reformed liberalism that recognized and relied upon important traditional foundations for social order:

> Instead of accepting this joint view of orthodox Liberals and collectivists, I consider that the alternative to the planning of cultural and economic life is not some inconceivable system of absolute *laissez faire* in which the State is supposed to wither away, but that the alternative is freedom under law and custom as laid down, and amended when necessary by the State and public opinion. It is law, custom, and public opinion which ought to govern society in such a way that by the guidance of their principles the energies of individual exertions are sustained and limited.
>
> (Polanyi 1940/1975, 59)

Polanyi was convinced that modern society and culture, shaped by science, could best prosper in a politics focused on individuals who could pursue personal projects that promoted the common good: "civilization consists mainly in the system of behavior by the observance of which men and women will benefit rather than injure their fellows while pursuing their own personal interests in life" (Polanyi 1940/1975, 59–60).

2.5 Deeper Roots of Social Order

Polanyi was clearly attuned to what he regarded as some deeper roots of social order that underlay his political philosophy mediating between the poles of "extreme liberalism" and "totalitarianism".[6] What he called "moral confidence" was certainly one of those deeper roots:

> [W]ithout moral confidence between men, there can be no government by the consent of the governed. ... Thus inevitably, once we deny that moral motives play a part in politics, we find that the only possible form of government is by force.
>
> (Polanyi 1947b, 11–12)

Polanyi thought pervasive modern "moral skepticism" threatened "the very foundations of man's communal life" (1947b, 9; see discussion below) and he affirmed that "a free society can exist only if men firmly believe in each other as essentially moral beings" (1947b, 11). If "free government is guided by discussion", he asked, "how can you argue with people who have no moral conscience?" (1947b, 11).

Polanyi seems to have been keenly aware of political minefields with an epistemic dimension that he saw in the political discourse in the European societies he knew in the middle decades of the last century. He paid attention to the importance of trust:

> The widely extended network of mutual trust, on which the factual consensus of a free society depends, is fragile. Any conflict which sharply divides people will tend to destroy this mutual trust and make universal agreement on facts bearing on the conflict difficult to achieve.
>
> (Polanyi 1956, 16)

He was very wary about the revolutionary enthusiasm in his era since it undercut the grounds of factuality in society and undermined rational political processes and compromise:

> a modern revolutionary government aiming at the total renewal of society, inevitably destroys the consensus of trust that underlies the process of fact-finding by severing all ties with its opponents. Whoever is not its unconditional supporter is held to be its mortal enemy.
>
> (Polanyi 1956, 16)

Populism of all types Polanyi regarded as dangerous because it covered over differences and downplayed the importance of freedom as central to reasonable governance processes. In 1936 in "Truth and Propaganda", a discussion treating the Webbs' account of Soviet Communism, Polanyi (1940/1975, 96) declared, "Freedom to-day [sic] is drowned in popular emotion". But it was not only the rise of popularism that Polanyi feared in political cultures; he also noted the kind of unrelenting polarization that seems common in modern political cultures, creating a kind of intractability that makes for social fragility. Polanyi linked his emphasis upon the importance of freedom and the inevitability of change in modern societies with a call for a more irenic disposition:

> The fight for freedom must aim … centrally at the voluntary reunion of conflicting groups. This is the unending task of those dedicated to the service of liberty. For life in a changing society can never cease to produce new dissensions and free citizens can therefore never pause in their search for new harmonious solutions to ever recurring conflicts.
>
> (Polanyi 1947c, 1058)

2.6 Polanyi's General Models of Modern Social Order

Particularly in his early writing, Polanyi used a dualistic general framework for discussing social organization. He often sharply contrasted what he called "planning" (or "corporate orders" that are "centrally directed" [Polanyi 1951, 112–114]) with "supervision" (or systems of "dynamic order" or "spontaneous order" [Polanyi 1951, 114–122, 154]):

> [T]he essence of planning is the absorption of the actions held under control by a single comprehensive scheme imposed from above. It is the co-ordination of these actions by means of vertical lines of authority which impose a specific task on each subordinate unit.
>
> (Polanyi 1940/1975, 35)

Both communists and fascists (and those influenced by them) were infatuated, Polanyi believed, with "planning", an approach that presumes it is possible comprehensively to know and shape the future in a complex, dynamic modern society.

Polanyi contended that totalitarian states address complexity with bureaucratic, centralizing strategies. They hold that "social order is upheld by the commands of the State" and that independent actions of individuals "can never claim to perform a social function that positively contributes to order" (Polanyi 1941, 429–430). Only the state can represent and enhance the collective interests of the community. Moreover, personal individual action is an exercise of private freedom that is irrelevant to the common good and usually must be curtailed for the common good. Polanyi believed such assumptions about the state (i.e., that it is solely responsible for collective interests) and liberty (i.e., that it is solely concerned with individual interests) had, by the thirties, become common in British society, and this was largely a reaction to the problems created by extreme liberalism as well as the rise of totalitarian states. He suggested that British ideas about individual freedom, growing out of extreme liberalism, brought "contempt on the name of freedom", and thus, progressive citizens had come to believe that only some form of totalitarianism could "be the sole guardian of social interests" (Polanyi 1940/1975, 58).

Polanyi countered this ethos by articulating a more restricted role of the state and broader ideas about freedom. He believed a reformed liberalism could avoid the excesses and disasters that he saw in contemporary communist and fascist states. He was convinced that individual personal actions do produce public benefit. Consequently, the state should facilitate this with laws and policies protecting individuals and promoting their responsible actions in important specialized intellectual communities such as science and the law, as well as in the broader, largely competitive market economy.

Polanyi carefully monitored Soviet programs to centralize the economy, the most extensive contemporary experiment in economic planning, and he did intensive research on the Soviet economy from the period of the Russian Revolution forward.[7] He made several scientific trips to the Soviet Union between 1928 and the mid-thirties and this stimulated his interest in the Soviet economy. He three times published an early statistical study focusing on the floundering Soviet economy (Polanyi 1935, 1936, and 1940/1975, 61–96), and he argued that Stalin-era improvements in the Soviet economy were achieved by incorporating market mechanisms, though these were not acknowledged (Polanyi 1937b, 29–30). Polanyi contended that complex modern industrial societies shaped by sophisticated science and technology must incorporate market mechanisms in order to function reasonably effectively (Polanyi 1951, 138, 160–167). Some Polanyi criticisms of centralization of economies focus on the failure of such systems to properly estimate demand (Polanyi 1951, 138–153).

Polanyi was most directly involved in opposing the planning project that aimed to centralize scientific research and development in Britain. In the early forties, Polanyi became a leader in the newly established Society for Freedom in Science, and he dated his own turn from science to philosophy to an early 1941 meeting of this group (Scott and Moleski 2005, 184). Several Polanyi essays such as "The Planning of Science" (Polanyi 1946b) and "Pure and Applied Science and Their Appropriate Forms of Organization" (Polanyi 1953) appeared in the Society's occasional pamphlet series.[8] Polanyi argued centralization would seriously impede modern science and this would also be detrimental for modern society for which science increasingly provided an important intellectual framework and (more indirectly) contributions to material culture.[9] He sharply distinguished pure scientific research from applied scientific research that aimed to produce ends currently viewed as socially important (Polanyi 1953). Both pure research and applied science and technological development, of course, remain important today. Polanyi claimed it is imperative to recognize differences between discovery-oriented and practical endeavors (i.e., the fruit of discovery-oriented science) and to cultivate both pure and applied science. However, this cultivation must take place in different socio-economic contexts.[10]

Polanyi countered the philosophical and economic emphasis upon planning in societies in the thirties and forties by emphasizing "regulative or supervisory authority" (Polanyi 1940/1975, 36), which he understood as an alternative organizational strategy eschewing "the absorption of ... actions held under control by a single comprehensive scheme imposed from above" (1940/1975, 35). "Supervision" relies on the cultivation of ideas and practices operating in the many subcultural niches of society.[11] Such niches Polanyi called "centres" in 1940 (see quotation below) and, after he discovered and adapted Wolfgang Köhler's ideas, "dynamic" orders (Polanyi 1941, 435) and eventually "spontaneous" orders (Polanyi 1951, 154).[12] He envisioned society as a complex overlapping network of such subcultural social orders; these communities of interpretation and common practice are not necessarily identical but have important family resemblances and they have some impact on each other (Mullins 2013). These social orders include specialized intellectual domains like science (with many different overlapping neighborhoods) and the law, but they also include non-intellectual practical domains concerned with production and distribution, as well as cultural domains concerned with activities focused on art and religion:

> Supervision presupposes human activities which are initiated from a great multitude of centres, and it aims at regulating these manifold impulses in conformity with their inherent purpose ... [S]uch functions can be exercised only in a Liberal society to which the cultivation of widely dispersed sources of initiative is essential and ... mental communications are open throughout the community. Public supervisory powers are in fact the vital safeguards of independent forces of initiative in society, the integrity of which they are to protect against private corruption as well as against oppression by collectivist tendencies of the State.
>
> (Polanyi 1940/1975, 36)

Although he discussed basic distinctions between intellectual and non-intellectual social orders (Polanyi 1951, 162–167), Polanyi argued more generally that public authority should make "generally available social machinery and other regulated opportunities for independent action" and should allow all individual agents to "interact through a medium of freely circulating ideas and information" (Polanyi 1940/1975, 36):

> When order is achieved among human beings by allowing them to interact with each other on their own initiative – subject only to laws which uniformly apply to all of them – we have a system of spontaneous order in society. ... [T]his self-coordination justifies their liberty on public grounds.
>
> (Polanyi 1951, 195)

Many important social tasks in a society are carried out by persons acting in the context of relatively autonomous social networks and, within such orders, socially aligned persons already interact with a certain amount of independence and discretion. Persons learn the practices and ideals of a social network as apprentices whose development is supervised by experienced practitioners. The activities of most members of a dynamic order are coordinated without much need for external direction because members have common guiding principles and values (or transcendent ideals such as concern for truth in science and justice in legal affairs) that promote social coherence and ongoing public conversation and the self-transformation of the network.

Even before he worked out more carefully his epistemological ideas in the fifties, Polanyi insisted that there is a deep connection between "understanding, believing and belonging ... three aspects of the same state of mind", and that this connection is manifestly visible in science, in religion, and in civic morality (Polanyi 1947b, 6). Especially within intellectual orders, but also other social orders, traditions are important and can become the basis for cultivation of new ideas and skills. Respect for innovation can be nurtured so that such orders are dynamic.[13] In a 1962 essay, Polanyi noted the "dynamic orthodoxy" of science and other social orders that constitute a free society; such orders strive for self-improvement (Polanyi 1969, 70).[14] Polanyi sought to facilitate and to promote a better appreciation for the growth of thought in society (see discussion below).[15] He thus regarded the exercise of supervisory authority as the key to a stable but changing society in which thought grows. In a stable but changing society, tradition and change are bound together as "dynamic orthodoxy" in a context in which authority is not altogether political and single as in totalitarian society.

2.7 Democracy, Law, and Public Liberty

Polanyi was committed to democratic government, but he was clear that democracy must be based on principles: "Freedom is government by principles ... Democratic self-government is the appeal of interested parties to public opinion on the basis of common principles. Unprincipled democracy is lawless and its dominion is tyrannical" (Polanyi 1945a, 1). He tightly linked law, universal suffrage, and what he called "public liberty" as the framework of a democratic order:

> We may call democratic a political system in which the rule of law sustains public liberties under a government elected by universal suffrage. ... [P]ublic liberties are the heart of democracy; the rule of laws is its muscular framework; and finally, a democratically elected government forms a dynamic centre for improving the laws by which men live in a free society.
>
> (Polanyi 1958, 17)

Public liberties are "the heart of democracy" because they promote, in the context of law and universal suffrage, the participation of individuals in the many subcultural social orders of complex modern society. As the discussion above has implied, Polanyi argued that "the logic of public liberty is to co-ordinate independent individual actions spontaneously in the service of certain tasks" (Polanyi 1951, 198).

Polanyi sharply distinguished "public" and "private" liberty because he believed that a reformed liberal, democratic society must better understand, protect, and promote public liberty. Because modern societies are complex, highly differentiated social entities, they require "a more general view of liberty, extending beyond the claims of private freedom" (Polanyi 1941, 431). Although early modern democratic experiments focused on individual freedoms and their protection, Polanyi claimed such a focus was insufficient. He rejected "the individualistic formula of liberty" that once was taken to be self-evident, pointing out that this view could be "upheld only in the innocence of eighteenth century rationalism" (Polanyi 1951, vii). "The freedom of the individual to do as he pleases, so long as he respects the other fellow's right to do likewise", Polanyi proclaimed, "plays only a minor part" in his "theory of freedom" – "private individualism" is not "an important pillar of public liberty" (Polanyi 1951, vii).

Polanyi tied a "free society" to the cultivation of and dedication to "a distinctive set of beliefs", and public liberty is a social device used to nurture and protect such cultivation and dedication. Public liberty is thus a positive conception of freedom and is "not for the sake of the individual at all, but for the benefit of the community in which dynamic systems of order are to be maintained" (Polanyi 1941, 438). Public liberty is "freedom with a responsible purpose: a privilege combined with duties" (Polanyi 1941, 438).

Public liberty thus promotes responsible individual action in a subcultural community such as science that, in turn, ultimately helps produce the growth of scientific thought that Polanyi took to be a common good. Law and policy can facilitate the responsible scientist's or judge's effort to fulfill obligations or duties. Duties are always situated in a certain social niche or dynamic order such as the scientific or legal community. The responsible scientist takes on research projects and participates in the program for evaluating the research of other scientists, and this participation in the public conversation of scientists slowly contributes to the development of a richer scientific perspective.[16] In an analogous way, judges exercise public liberty. They have an orderly process for receiving and considering cases in terms of precedents, and through their diligence, society works to realize the ideal of justice. Law and democratic decision-making processes (which can reform law and policy-making processes) Polanyi thus emphasized must structure and foster the positive freedoms promoting the dutiful action of individuals with vocations in social enclaves such as that of science, law, and the economy.

To summarize Polanyi's reformed liberal agenda, he sought to emphasize in government matters rooted in democratic principles and practices, law, and public liberty, but Polanyi stressed that these elements must be underlain by trust and moral confidence. Moral confidence provides the seedbed in which there can grow a consensus about facts and a relatively high consensus about facts is important in a stable but dynamic society.

Today, of course, in emerging twenty-first-century digital culture, it is particularly difficult to maintain the kind of relative consensus about facts that is important if the transitions in society are to be orderly and gradual. Across the world, many contemporary politicians do not recognize the importance of ideas about truth and its independent pursuit, and they manipulate democratic principles and practices and law. Some politicians massively tweet and thrive on chaos; many seem to spend enormous energy on "messaging" designed to cultivate division in emerging digital culture. Some politicians also actively undermine confidence in science and in the independent public media; some simply control the public media. In sum, it seems many contemporary politicians in emerging digital culture are willing to ignore the deeper roots of science and society that Michael Polanyi spent much of his life in the thirties and forties trying to identify and comment upon.

2.8 Problems of the Modern Mind: The Coupling of Skepticism and Moral Passion

Although Michael Polanyi, in the thirties and forties, frequently wrote about social organization and political philosophy, his interests also reached into the history of science and the history of ideas.[17] He came to be keenly attuned to the way science has been represented in modernity and the impacts of this representation.[18] Polanyi firmly believed ideas matter – they decisively shape human affairs. In particular, Polanyi criticized the ways "scientism"[19] eventually became dominant in the natural sciences and this bled into the social sciences and the modern Western cultural mainstream and created what Polanyi regarded as the problems of the modern mind.

By mid-century, Polanyi saw that his agenda reforming liberalism needed to be situated in the context of a wider cultural criticism and a constructive philosophical perspective aimed at redirecting modern culture shaped by science. An important component of this wider cultural criticism is Polanyi's complex discussion of the problematic dispositions of the modern mind that are, according to Polanyi, the outgrowth of the way science has been mis-understood and misrepresented in modernity. In his middle and late thought, Polanyi described and deconstructed these destructive dispositions, replacing them with a new, richer understanding of human beings rooted in his discovery-centered account of science and his innovative account of human knowing that appreciated its personal and social roots.

The mainstream narrative about the developing modern scientific tradi-tion, according to Polanyi, is a misleading narrative focused on method, on providing reductionistic, mechanistic, and materialistic explanation,[20] and, pre-eminently, on objectivity. As science evolved after the scientific revolu-tion, it came to be represented as a deterministic and materialistic view from nowhere. This is a narrative, in Polanyi's view, not attentive to scientific discovery and it does not have a deep understanding of the operation of the scientific community.[21] Polanyi's post-critical account of science, which focused around his ideas about personal knowledge, challenges these ele-ments in the mainstream account (see discussion in the next section).

Polanyi's cultural criticism argued that elements of the mainstream nar-rative about science have come to be united with the increase (and secular-ization) of moral passions in persons living in modern societies after the French Revolution (PK, 213–214; Polanyi 1965, 12–14).[22] Modern society, drawing on scientific, technological, and economic developments, has brought important improvements of various kinds to modern life.[23] There has been a revolution of expectations in modernity that has often been invested by modern people in utopian ideas about the transformation of society. This rise in moral passions came to be, in the twentieth century, combined with the scientistic narrative focused on doubt as a method, materialism, and objectivism. As Polanyi put matters (in what he dubbed "a strange story"), "the two conflicting ideas of our age – its skepticism and its moral passions – are locked in a curious struggle in which they may combine and reinforce each other" (Polanyi 1965, 12).

This combination has decisively shaped the modern mind which Polanyi carefully characterized "as a body of ideas having their origin in thought" (Polanyi 1965, 12). This description shows Polanyi's resistance to modern reductionist accounts of mind to things more tangible, which, of course, goes hand in hand with undervaluing the influence of thought in modern human affairs.[24] Polanyi most straightforwardly and provocatively puts his thesis about the union of conflicting dispositions of the modern mind in an essay republished in *The Logic of Liberty*:

> A new destructive skepticism is linked ... [in the modern world] to a new passionate social conscience; an utter disbelief in the spirit of man is coupled with extravagant moral demands. We see at work here the form of action which has already dealt so many shattering blows to the modern world: the chisel of skepticism driven by the hammer of social passion.
>
> (Polanyi 1951, 4)[25]

In Polanyi's account, the chisel of skepticism driven by the hammer of excessive moral passions led to the nightmare of violence, nihilism, and totalitarianism in the twentieth century. Skepticism and moral passion undercut earlier beliefs in the reality of key social values such as the existence of independent truth, the importance of reasoned controversy, and acceptance of ideals such as compassion and justice. Modern people locate ultimate reality in such forces as power, economic interest, and subconscious desires. Polanyi thus insistently advocated that we must develop new habits of thought that will overcome our tendency to allow the chisel of skepticism to be hammered by our moral passions in a way that destroys the fabric of public and private life.

Polanyi sometimes pointed to the "dynamo objective coupling" (PK, 230) that he claimed operates in the minds of many who live in modern societies. Burgeoning modern moral passions (directed toward improvements in society) are undercut by the excessively critical modern temperament cultivated by scientism; this produces a dynamic coupling whose impact Polanyi traced not only to manifestations in individual modern consciousness (leading to violence and nihilism, which was first explored in modern literature) but also to social policies (some horrifying) of modern totalitarian regimes. This coupling sets up "self-confirmatory reverberations". Purportedly scientific claims (such as those made in Marxian ideas about history and society) are accepted because they "satisfy moral passions" and further excite such passions "and thus lend increased convincing power to the scientific affirmations in question – and so on indefinitely". But the coupling is "potent in its own defence" since moral passion deflects criticism of the purportedly scientific affirmations and moral objections are "coldly brushed aside" as unscientific (PK, 230).

The term that Polanyi most frequently used to summarize his analysis of the problems of the modern mind is "moral inversion". He characterized modern human beings as readily susceptible to inversions that are in fact both an individual and cultural-historical process of deformation. D.M. Yeager has carefully analyzed Polanyi's many discussions of "moral inversion" and its impacts, showing how inversion fits into the general framework of Polanyi's evolving convictions about modern intellectual history (briefly outlined above):

> "moral inversion" ... may be broadly understood as the process by which the fusion of scientific skepticism ... with utopian social aspirations ... produces the dystopia of moral and political nihilism out of which arises the modern totalitarian state, in which the only principle of social order is absolute coercive power and in which material welfare is embraced as the supreme social good.
>
> (Yeager 2002–2003, 23)[26]

Polanyi contended "the morally inverted person has not merely performed a philosophic substitution of moral aims by material purposes, but is acting with the whole force of his homeless moral passions within a materialistic framework of purposes" (Polanyi 1951, 106). This produces some of the odd paradoxes of modernity such as the "moral appeal of immorality" (PK, 232) and "the moral appeal of a declared contempt for moral scruples" (PK, 235). Yeager's broader discussion suggests both the complexity and the subtly of Polanyi's many discussions of "moral inversion" from the mid-forties until the end of his life:

> Moral inversion is not one thing, but a cascade of paradoxical inversions that break upon us. Taken together, they have the potential to bring to an end the culture and civilization that arose out of the Renaissance and Enlightenment in Europe. In the name of social well-being, society is immeasurably impoverished. In the name of social justice, justice is trampled upon. In the name of self-determination, self-determination is denied. In the name of freedom, freedom is lost. In the name of morality, immorality is celebrated. In the name of truth, the possibility of arriving at truth is denied. In the name of the highest moral aspirations, the West descends into "soul-destroying tyrannies" (PK, 265). Liberalism is devoured by her own children, and the Enlightenment is by the Enlightenment destroyed.
>
> (Yeager 2002–2003, 34)

Polanyi's complex criticism of modernity's inversions is an account in which "the good is lost through complicated, misguided, and unrealistic dedication to the good" (Yeager 2002–2003, 22). I suspect today, two decades into the twenty-first century, that the chisel of skepticism hammered by extraordinary moral passions remains a driving force in modern societies, one which stokes populism and forms of nationalism that undermine the modest successes in international cooperation achieved after the middle of the twentieth century.

2.9 Re-conceiving Knowing, Scientific Knowledge, and Scientific Practice

By the middle and late forties, Polanyi began to work out an alternative narrative (about science) to what he believed had become powerfully dominant and culturally destructive in his lifetime. Re-conceiving scientific practice and scientific knowledge and ultimately all human knowing became the ambitious constructive philosophical project to which Polanyi devoted the last 30 years of his life. In a 1948 essay, Polanyi asserted that we must examine the "foundations of modern thought" and "realise at last that skepticism cannot in itself ever discover anything new". Skepticism can release "powers of discovery, but the powers must always spring from belief" (Polanyi 1948, 100). For Polanyi, discovery is the central puzzling foundation of modern science, but scientists and philosophers of science have paid little attention to this foundation. Polanyi linked discovery in science to belief and ordinary human perception, and thus discovery became the paradigm case of human knowing (SFS, 21–25, 31–38).

Rather than focusing on scientific method (or verification and falsification) and the so-called objectivity and impersonality of scientific inquiry and scientific knowledge, Polanyi emphasized the skillful nature of scientific inquiry (which includes the formulation of good problems) and the extraordinary skills of the master scientist who makes discoveries. Special skills and conceptual tools are cultivated in a specialized community and a practicing scientist working in such a community acquires certain beliefs and a form of specialized perception. However, Polanyi emphasized the continuity between ordinary perception and the specialized perception of the scientist who first sees and then solves scientific problems. He also stressed, given his emphasis on the sophisticated skills of a scientist, that much about scientific knowledge and any other kind of knowledge remains unspecifiable. All knowers are actively engaged participants in acts of comprehension. They dwell in certain elements of which they are subsidiarily aware (i.e., tacitly held elements) and integrate such elements to attend to what is of focal interest. As Marjorie Grene put it, this anti-Cartesian Polanyian approach focusing on "tacit knowing" provides "grounds for a revolution in philosophy":[27]

I regard knowing as an active comprehension of the things known, an action performed by subordinating a set of particulars, as clues or tools,

to the shaping of a skillful achievement, whether practical or theoretical.
We may then be said to become "subsidiarily aware" of these particulars
within our "focal awareness" of the coherent entity that we achieve.

(PK, xiii)

This is Polanyi's account of the profoundly "personal" nature of knowing
that involves the "participation" of the knower in shaping knowledge and
that produces "personal knowledge":

The participation of the knower in shaping his knowledge, which had
hitherto been tolerated only as a flaw – a shortcoming to be eliminated
from perfect knowledge – is now recognized as the true guide and
master of our cognitive powers. ... We must learn to accept as our ideal
a knowledge that is manifestly personal.

(Polanyi 1959, 26–27)[28]

Polanyi emphasized that inquiry is a peculiarly human vocation always
infused with intellectual passions. Such passions drive the quest for scientific
discovery and for ordinary understanding. Polanyi called his account of scientific inquiry, and, more generally, his description of human knowing, his
"fiduciary" philosophy (PK, ix, 264–268):[29]

We must now recognize belief once more as the source of all knowledge. Tacit assent and intellectual passions, the sharing of an idiom and
of a cultural heritage, affiliation to a like-minded community: such are
the impulses which shape our vision of the nature of things on which
we rely for our mastery of things. No intelligence, however critical or
original, can operate outside such a fiduciary framework.

(PK, 266)

Scientists and indeed all knowers, Polanyi contended, trustingly rely on
tacit knowledge as that from which they attend to matters of direct interest.
All knowledge thus has a from–to structure. The person's fund of subsidiary
knowledge is learned in trusting interaction with others in a like-minded
community. Although scientists and all human beings acquire and hold what
Polanyi dubbed "personal knowledge", such knowledge is not merely whim-
sically subjective. "Personal knowledge" is backed by commitment and held
with what Polanyi called "universal intent" (PK, 65, 300–306, 308–316, 377–
379). That is, it is held in such a way that one believes that other inquirers
who make contact with reality will arrive at the same conclusion.

2.10 Polanyi's Comprehensive Vision of Science and Society

In his constructive post-critical philosophy, Polanyi articulated a comprehensive vision of science and society. He outlined the way in which the

scientific enterprise might become leaven in modern societies. Insofar as science looks toward the unknown, it can stimulate other social endeavors and help broadly contour modern societies and cultures, orienting them toward change and pursuit of transcendent ideals. Science is an interesting, growing "organism of ideas" (Polanyi 1940/1975, 4) constantly reaching into the future. Scientists dwell in the current scientific tradition in order to break out, to move beyond the present boundaries of scientific thought. It is this focus on the growth of thought through ongoing inquiry that Polanyi sought to recover as central in modernity:

> Scientific tradition derives its capacity for self-renewal from its belief in the presence of a hidden reality, of which current science is one aspect, while other aspects of it are to be revealed by future discoveries. Any tradition fostering the progress of thought must have this intention: to teach its current ideas as stages leading on to unknown truths which, when discovered, might dissent from the very teaching which engendered them. Such a tradition assures the independence of its followers by transmitting the conviction that thought has intrinsic powers, to be evoked in men's minds by intimations of hidden truths.
>
> (Polanyi 1966, 82)

Polanyi argued that scientists make contact with reality, but reality remains always partially hidden. Current scientific discoveries have indeterminate future manifestations and thus current discoveries appeal to future discoveries as that which will reveal more richly what scientists now only vaguely recognize. Tradition is taught for the purpose of inspiring novices to reform the scientific tradition as they see more deeply into the nature of reality. It is this curious dynamic of self-renewal, produced by relying on respect for tradition and the believed-in immense richness of reality, that Polanyi thought it most important for those in scientific professions as well as modern politicians and ordinary citizens to appreciate. Without embarrassment, Polanyi affirmed the importance of thought and envisioned a "Society of Explorers":

> It is the image of humanity immersed in potential thought that I find revealing for the problems of our day. It rids us of the absurdity of absolute self-determination, yet offers each of us the chance of creative originality, within the fragmentary area which circumscribes our calling. It provides us with the metaphysical grounds and the organizing principles of a Society of Explorers.
>
> (Polanyi 1966, 91)

Clearly, Polanyi sought to extend the ideal of a "Society of Explorers" beyond the scientific community to all areas of modern society. He argued that "the whole purpose of society lies in enabling its members to pursue their transcendent obligations". Achieving material well-being is "not the real purpose of society" but a "secondary task" that provides "an opportunity to fulfil ... true aims in the spiritual field" (SFS, 83). As Polanyi later put matters, human beings "need a purpose which bears on eternity. Truth does that; our ideals do it" (Polanyi 1966, 92). Polanyi thus characterized the required broad purpose in a "Society of Explorers" in terms of a return to the open acceptance of human ideals like truth, justice, and charity that were more directly embraced in early modern society. Polanyi is not, however, a naive optimist about modern society. He firmly believed that modern human beings must seriously work to develop a deeper understanding of science, one that will eliminate the scientism so prevalent in modern society. But he makes it clear that a return to earlier ideals can sustain a "Society of Explorers" only if modern people can also learn to "be satisfied with our manifest moral shortcomings and with a society which has such shortcomings fatally involved in its workings" (Polanyi 1966, 92). At the heart of Polanyi's thought about science and modern society, there is both optimism and a blunt sobriety learned from recent history:

> We must somehow learn to understand and so to tolerate – not destroy – the free society. It is the only political engine yet devised that frees us to move in the direction of continually richer and fuller meanings, i.e., to expand limitlessly the firmament of values under which we dwell and which alone makes the brief span of our mortal existence truly meaningful for us through our pursuit of all those things that bear upon eternity.
>
> (Polanyi and Prosch 1975, 216)

Notes

1 For a discussion of "post-critical", see Mullins (2001; 2016) and Cannon (2016).

2 For a brief discussion of the historical development of Polanyi's philosophical ideas, including his account of his "fiduciary" program and the emerging "post-critical" era, see Mullins (2016, 3–6).

3 This section draws freely from the Polanyi biography, see Scott and Moleski (2005). See also Polanyi's brief autobiographical statement (Polanyi 1975), and the Royal Society biographical memoir (Wigner and Hodgkin 1977).

4 The 1964 Torchbook Edition, which contains the Torchbook Preface, is cited parenthetically hereafter as PK (see Polanyi 1958/1964).

5 From 1953 until 1968, Polanyi was deeply involved in the programs of the Congress for Cultural Freedom and continued some of his earlier work on questions about science and political freedom in the Cold War years.

6 In *Personal Knowledge* (published in 1958, after Polanyi has long worked on questions about social order), Polanyi discusses "four coefficients of societal organization": "the *sharing of convictions* ... the *sharing of a fellowship*, ... *co-operation*, ... [and] the exercise of *authority or coercion*". These can together "form stable features in the shape of social institutions". In modern society, which is "based on elaborate articulate systems and a high degree of

specialization", certain kinds of institutions and practices emphasize one or another of these coefficients (PK, 212).

7 Polanyi's account focuses on much more than simply the ongoing Soviet experiment (and disaster) with a centralized economy, although he certainly examines details of this carefully. Polanyi fits the Russian Revolution and subsequent events in Soviet history into a much broader narrative about the development of certain ideas after the scientific revolution and the culmination of these ideas in the modern crisis that is reflected in totalitarianism, nihilism, and violence in the twentieth century. As I note below, Polanyi's broader narrative is bound up not only with history, politics, and economics (including the rise of Marxian ideas), but draws also on the history of modern science and the ways in which developing modern science is interpreted and this shapes the development of modern ideas.

8 Polanyi essays, like these in this occasional pamphlet series, were published in more than one venue. Often Polanyi essays in the same period are similar and some seem to be revised versions of earlier essays. Polanyi published many things in a variety of journals, but anyone familiar with Polanyi's activities and friends notices certain patterns. Polanyi's several essays in the late forties and early fifties in the *Bulletin of Atomic Scientists* (e.g., Polanyi 1946a, 1949b), owe something to his growing friendship with Eduard Shils, who was affiliated with the *Bulletin*. Publications after 1953 in journals subsidized by the Congress for Cultural Freedom reflect Polanyi's involvement in Congress programs promoting freedom in science and his leadership role in Congress conferences and seminars.

9 Polanyi's account of science, as I discuss below, focuses on the centrality of discovery. He never believed that planning research was practical, and he objected to the idea that a strategy for research should fit into current social demands. But he did hold that eventually what pure scientific research produced filters into applied science and technology, and this produces a social impact.

10 Polanyi saw applied science and technology projects as tied to economic factors, but basic scientific research is not oriented to practical applications and he supported public funding for so-called "pure science". This view is akin to Polanyi's ideas about the importance of "public liberty" (see discussion below). Polanyi often seems to have regarded those who favored planned science as failing to appreciate basic research and this view likely was in part suggested to him by comments of figures like theoreticians such as Bukharin (SFS, 8).

11 In the late forties, Polanyi somewhat repackaged his early ideas about supervisory authority (operating in the many subcultural social networks in society) in terms of "polycentricity", which he works out primarily in his discussions of economics. Complex modern societies, he argued, must address the manageability of questions in a "polycentric order" (Polanyi 1951, 170–184). In two essays (that subsequently became chapters in Polanyi 1951, 111–137, 154–201), Polanyi developed mathematical arguments that he contended made his case "in semi-quantitative terms" (1951, 126). One essay argues "the span of control" in "a system of mutual adjustment" is larger than "under the authority of a corporate body" (1951, 115). That is, "the impossibility of central direction" lies in its short "span of control" compared with a self-adjusting system (1951, 126). Polanyi held that administering modern industrial production "requires the readjustment of a number of relations far exceeding the span of control of a corporate body" (1951, 115) and thus, the rational administration of modern industrial production calls for "spontaneous systems" (1951, 116). The second essay (1951, 154–200), Polanyi identified as drawing conclusions "more general" than those focusing on "the span of control" and as concerned with "the polycentric nature of the economic task" (1951, 184). He argued that some polycentric tasks that can be mathematically formalized can be "computed *exactly*", while others can

be computed "*by a sequence of successive approximations*" (and he outlined this method [1951, 173–176]), and a third group remain "altogether *incomputable*" (1951, 171). As noted above, Polanyi was a sophisticated economist who made an economics education film, wrote a primer on Keynesian economics, *Full Employment and Free Trade* (1945b), and wrote many articles addressing questions in economics. This work is integrally linked with his political philosophy, his philosophy of science, and even his later epistemology.

12　Mullins (2010) discusses Polanyi's extensive and complicated adaptation of Kohler's ideas in his early social and political philosophy and his later epistemology.

13　Polanyi regarded economic orders of production and distribution as more strictly competitive and as reliant on pricing with the medium of money. But this also produces innovations. Interestingly, Polanyi criticized laissez-faire views that assume there is only one "economic optimum" that the market can achieve. Invoking Dickens, he affirmed the nineteenth century was a time of "continuous social reform" and that "there exists an indefinite range of relative optima toward which a market economy can tend" (Polanyi 1951, 187; see also Polanyi 1946c).

14　Near the end of "Conviviality", the chapter in *Personal Knowledge* that is directly focused on social organization, Polanyi asks, "[c]an the beliefs of liberalism no longer believed to be self-evident, be upheld henceforth in the form of an orthodoxy? Can we face the fact that, no matter how liberal a free society may be, it is also profoundly conservative?" And he answers in the affirmative: "The recognition granted in a free society to the independent growth of science, art, and morality, involves a dedication of society to the fostering of a specific tradition of thought, transmitted and cultivated by a particular group of authoritative specialists, perpetuating themselves by co-option. To uphold the independence of thought implemented by such a society is to subscribe to a kind of orthodoxy" (PK, 244).

15　Polanyi's earliest theoretical account of society is a 1941 review article rebutting views presented in J.G. Crowther's *Social Relations of Science*; his article is titled "The Growth of Thought in Society" (Polanyi 1941) and Polanyi makes clear that science is one, but not the only domain, of modern thought. Much later, Polanyi also published one of his 1965 Wesleyan Lectures under the title "The Growth of Science in Society" (Polanyi 1967) and he clearly presents science as an important domain of modern thought. In writing reaching back to the late thirties, Polanyi suggests science is a growing "organism of ideas" (Polanyi 1940/1975, 4) developed in modernity, one with many connected branches and that has, for many modern people, order and cohesion and a "supreme attraction for the human mind" (Polanyi 1940/1975, 5).

16　Polanyi's niche-specific notions of freedom and responsibilities (i.e., his emphasis on public liberty) seem narrow at times since human beings usually participate in a set of social niches as well as, one might argue, society at large.

17　Polanyi articulates, in both early and late writing, a nuanced and complex account of the historical development of primary ideas in physics and chemistry. For example, even in a short essay, he outlines how Pythagorean underpinnings of views like those of Kepler gave way to ideas about a mechanical universe that were modified by field theories and then replaced by relativity that was further modified by "a purely statistical interpretation of atomic interaction" (Polanyi 1947a, 13).

18　By the mid-forties Polanyi's essays show that he already was probing issues concerned with the different ways Enlightenment ideas had developed and been assimilated in Britain and on the Continent and how this assimilation was reflected in political and cultural history. "Science and the Modern Crisis" contrasts the history and the development of ideas on the Continent with what Polanyi calls the British suspension of the "logic of the Leviathan" (Polanyi

1945c, 116). British leaders did not link scientific materialism and progress: "religion retained a dominant position in the public life ... and moral arguments retained their position in the guidance of public policy" (Polanyi 1945c, 116). See also Polanyi (1943).

19 In correspondence from the mid-forties, Polanyi suggests he has borrowed this summary term (which he eventually uses a great deal) from Hayek (who may also have borrowed it). Polanyi and Hayek collaborated and corresponded in resisting "planned" science from the late thirties and Polanyi also likely saw some of the Hayek articles in the early forties in *Economica* that use the term. Polanyi uses "scientism" to designate reductionist, materialistic, mechanistic, and objectivistic accounts of science. He argued such views have been much more destructive in the social sciences than in the natural sciences. He particularly objects to the ways in which social scientists who promote such accounts present their views as what constitutes a scientific account. "Scientism" is closely aligned for Polanyi with "positivism" and "empiricism", terms he uses somewhat loosely. As I note below, Polanyi holds that "scientistic" views have come to shape the modern mind and this is the source of serious political and cultural problems.

20 Polanyi sharply criticizes reductionistic and atomistic views that construe scientific explanation as narrowly oriented to least parts, regarded as fundamental elements recognized in physics and chemistry. In writing in the sixties, he counters such a single-level ontology by working out a richer metaphysical account (i. e., a hierarchical ontology) that attends to the levels of reality and the ways in which boundaries between levels are controlled by a sequence of principles (Polanyi 1965, 216–218).

21 Polanyi particularly focuses on the negative impact of objectivism as a modern ideal of knowledge, and he counters objectivism with his account of personal knowledge. Laplace's vision, "the Laplacean fallacy" (PK, 142) is the poster child for all that is wrong with objectivism (PK, 139–142). Polanyi does carefully redefine "objective" and "subjective" and he avoids the common and careless bifurcated use of this terminology to characterize knowing and knowledge in his constructive account of personal knowledge.

22 Polanyi links and distinguishes intellectual and moral passions as important forces in modern life (PK, 216) and, at times, Polanyi also links the rise in moral passion to the secularization of Christian hope (Polanyi 1965, 12). After the American and French Revolutions, the idea gradually spread that "society could be improved indefinitely by the exercise of political will of the people, and that the people should therefore be sovereign, both in theory and fact". This movement gave birth to "modern dynamic societies" of two types, those in which the "dynamism is revolutionary" and those in which "dynamism is reformist". The former are modern totalitarian societies that subordinate "all thought to welfare" and the latter are modern free societies that "accept in principle the obligation to cultivate thought according to its inherent standards" (PK, 213).

23 Polanyi held "scientific skepticism smoothly cooperated at first with the new passions for social betterment" and this brought many important improvements in modern life. But by the twentieth century, a "sharpening of skepticism" led to "questioning the very existence of intangible things" (Polanyi 1965, 13).

24 Modern reductionistic objectivism influences people to "distrust intangible things" and to look behind intangibles "for tangible matters on which it relies for understanding" (Polanyi 1965, 12), and this is perhaps nowhere clearer than in contemporary discussions of mind and what lies behind thought. Polanyi (1965) addressed this directly in "On the Modern Mind".

25 Polanyi asked, "why moral forces could be thus perverted" in modernity and responded that the "great social passions of our time" became violent and destructive forces because "there was no other channel available to them". The

scientistic ethos emphasizing doubt destroyed "popular belief in the reality of justice and reason", identifying such ideas as "mere superstructures; as out-of-date ideologies of a bourgeois age; as mere screens for selfish interests hiding behind them; and indeed, as sources of confusion and weakness to anyone who would trust in them" (Polanyi 1951, 5).

26 As Yeager (2002–2003, 22) notes, "moral inversion" is a "preoccupying theme" for Polanyi, and he charts many rivulets of modernity's intellectual development. Yeager (2002–2003, 25–27) discusses a dozen modern intellectual trends that Polanyi treats as leading to inversion and totalitarianism and shows how Polanyi envisions their various combinations.

27 The title of Grene's article (published soon after Polanyi's death) evaluating Polanyi's philosophical contribution was "Tacit Knowing: Grounds for a Revolution in Philosophy". She noted that Polanyi's thought provided a "major break with the [philosophical] tradition and a possible foundation for a new turn in the theory of knowledge and, *a fortiori*, in philosophy as such" (Grene 1977, 164). See Mullins (1997; 2007) for a discussion of Polanyi's anti-Cartesian orientation in terms of his "participative realism" and his ideas about "comprehensive entities".

28 Polanyi carefully clarified his ideas about the "personal": "We may distinguish between the personal in us, which actively enters into our commitments, and our subjective states, in which we merely endure our feelings. This distinction establishes the conception of the *personal*, which is neither subjective nor objective. In so far as the personal submits to requirements acknowledged by itself as independent of itself, it is not subjective; but in so far as it is an action guided by individual passions, it is not objective either. It transcends the disjunction between subjective and objective" (PK, 300). See the discussion below.

29 "Fiduciary program" is terminology Polanyi uses in *Personal Knowledge* which he notes is focused on "the task of justifying the holding of unproven traditional beliefs" (PK, ix). But almost a decade before *Personal Knowledge*, Polanyi was emphasizing the centrality of belief in science. He argued science could not operate without the exercise of "fiduciary decisions" (1949a, 17). As Polanyi and others have noted, later developments in Polanyi thought are "less occupied with the justification of our ultimate commitments and concentrate instead on working out precisely the operation of tacit knowing" (PK, xi).

References

Bíró, G. 2020. *The Economic Thought of Michael Polanyi*. New York: Routledge.

Cannon, D.W. 2016. "Being Post-Critical." In *Recovering the Personal: The Philosophical Anthropology of William H. Poteat*, edited by D.W. Cannon and R.L. Hall, pp. 21–46. Lanham: Lexington Books.

Grene, M. 1977. "Tacit Knowing: Grounds for a Revolution in Philosophy." *Journal of the British Society for Phenomenology* 8 (3): 164–171.

Mullins, P. 1997. "Polanyi's Participative Realism." *Polanyiana* 6 (2): 5–21.

Mullins, P. 2001. "The Post-Critical Symbol and the 'Post-Critical' Elements of Polanyi's Thought." *Polanyiana* 10 (1–2): 77–90.

Mullins, P. 2007. "Comprehension and the 'Comprehensive Entity': Polanyi's Theory of Tacit Knowing and Its Metaphysical Implications." *Tradition and Discovery* 33 (3): 26–43.

Mullins, P. 2010. "Michael Polanyi's Use of Gestalt Psychology." In *Knowing and Being: Perspectives on the Philosophy of Michael Polanyi*, edited by Tihamér Margitay, pp. 10–29. Newcastle upon Tyne: Cambridge Scholars Publishing.

Mullins, P. 2013. "Michael Polanyi's Early Liberal Vision: Society as a Network of Dynamic Orders Reliant on Public Liberty." *Perspectives in Political Science* 42: 162–171.

Mullins, P. 2016. "Introduction to the Gifford Lectures." *Polanyi Society Web Site*. Available online at www.polanyisociety.org/Giffords/Intro-MP-Giffords-9-20-16.pdf. Accessed October 17, 2019.

Nye, M.J. 2011. *Michael Polanyi and His Generation: Origins of the Social Construction of Science*. Chicago: University of Chicago Press.

Polanyi, M. 1935. "USSR Economics – Fundamental Data, System and Spirit." *Manchester School of Economics and Social Studies* 16: 67–87.

Polanyi, M. 1936. *USSR Economics*. Manchester: Manchester University Press.

Polanyi, M. 1937a. "Congrès du Palais de la Découverte, International Meeting in Paris." *Nature* 140: 710.

Polanyi, M. 1937b. "Historical Society Lecture." Box 25, Folder 11, Michael Polanyi Papers. Special Collections. University of Chicago Library, Chicago, IL, pp. 1–32.

Polanyi, M. 1940/1975. *The Contempt of Freedom: The Russian Experiment and After*. London: Watts & Co (1940). New York: Arno Press.

Polanyi, M. 1941. [November] "The Growth of Thought in Society." *Economica* 8: 421–456.

Polanyi, M. 1943. "The English and the Continent." *Political Quarterly* 14: 372–381.

Polanyi, M. 1945a. [March]. "Civitas." Box 50, Folder 5, Michael Polanyi Papers. Special Collections. University of Chicago Library, Chicago, IL, pp. 1–7.

Polanyi, M. 1945b. *Full Employment and Free Trade*. Cambridge: Cambridge University Press.

Polanyi, M. 1945c. "Science and the Modern Crisis." *Memoirs and Proceedings of the Manchester Literary and Philosophical Society* 86 (6): 107–116.

Polanyi, M. 1946/1964. *Science, Faith and Society*. Chicago: University of Chicago Press.

Polanyi, M. 1946a. "The Foundations of Freedom in Science." *Bulletin of Atomic Scientists* 2 (11–12): 6–7.

Polanyi, M. 1946b. [February]. "The Planning of Science." *Publications of the Society for Freedom of Science*. Occasional Pamphlet No. 4.

Polanyi, M. 1946c. [April 13]. "Social Capitalism." *Time and Tide* 27: 341–342.

Polanyi, M. 1947a. [February]. "Science: Observation and Belief." *Humanitas* 1: 10–15.

Polanyi, M. 1947b. "What to Believe." Box 31, Folder 10, Michael Polanyi Papers. Special Collections. University of Chicago Library, Chicago, IL, pp. 1–13.

Polanyi, M. 1947c. [October 4]. "What Kind of Crisis." *Time and Tide* 4: 1056–1058.

Polanyi, M. 1948. "The Universities Today." *The Adelphi* 24: 98–101.

Polanyi, M. 1949a. "The Nature of Scientific Convictions." *19th Century and After* 146: 14–28.

Polanyi, M. 1949b. "Ought Science to be Planned. The Case for Individualism." *Bulletin of Atomic Scientists* 4 (1): 19–20.

Polanyi, M. 1951. *The Logic of Liberty: Reflections and Rejoinders*. London: Routledge and Kegan Paul.

Polanyi, M. 1953. [December]. "Pure and Applied Science and their Appropriate Forms of Organization." *Publications of the Society for Freedom of Science*. Occasional Pamphlet No. 14.

Polanyi, M. 1956. [November]. "The Magic of Marxism and the Next Stage of History." A Special Supplement of *The Bulletin of the Committee on Science and*

Freedom [Congress for Cultural Freedom Committee chaired by Michael Polanyi]. Manchester, UK.

Polanyi, M. 1958/1964. *Personal Knowledge: Towards a Post-Critical Philosophy.* New York: Harper and Row.

Polanyi, M. 1958. "Tyranny and Freedom, Ancient and Modern." *Quest* [Bombay]: 9–18.

Polanyi, M. 1959. *The Study of Man.* Chicago: University of Chicago Press.

Polanyi, M. 1965. "On the Modern Mind." *Encounter* 24: 12–20.

Polanyi, M. 1966. *The Tacit Dimension.* Garden City, NY: Doubleday.

Polanyi, M. 1967. "The Growth of Science in Society." *Minerva* 5: 533–545.

Polanyi, M. 1969. "The Republic of Science." In *Knowing and Being: Essays by Michael Polanyi,* edited by Marjorie Grene, pp. 49–72. Chicago: University of Chicago Press.

Polanyi, M. 1975. "Polanyi, Michael, (March 11, 1891–)." *World Authors, 1950–1970: A Companion Volume to Twentieth Century Authors,* 1151–1153. New York: Wilson.

Polanyi, M., and H. Prosch. 1975. *Meaning.* Chicago: University of Chicago Press.

Scott, W.T., and M.X. Moleski. 2005. *Michael Polanyi, Scientist and Philosopher.* New York: Oxford University Press.

Wigner, E.P., and R.A. Hodgkin. 1977. "Michael Polanyi 1891–1976." *Biographical Memoirs of Fellows of the Royal Society* 23: 413–448.

Yeager, D.M. 2002–2003. "Confronting the Minotaur: Moral Inversion and Polanyi's Moral Philosophy." *Tradition and Discovery* 29 (1): 22–48.

3 The Ethos of Science and Central Planning

Merton and Michael Polanyi on the Autonomy of Science

Péter Hartl

RESEARCH CENTRE FOR THE HUMANITIES, INSTITUTE OF PHILOSOPHY, MTA BTK
LENDÜLET MORALS AND SCIENCE RESEARCH GROUP

3.1 Introduction

This chapter presents and compares Robert K. Merton's and Michael Polanyi's criticism of totalitarian control of science, their defense of the autonomy of science, and their account of how freedom of science and a free society depend on each other. Coming from different intellectual backgrounds, both thinkers argue that a free society and scientific research are based on the same values. In the following, I will discuss Merton's and Polanyi's analyses of the role of the scientific community's epistemic and non-epistemic values and norms (referred to as the "ethos of science") in scientific research and free society. The main thesis of this chapter is that Polanyi's and Merton's arguments and ideas about academic freedom and totalitarian control of science have several points in common, even though Polanyi himself was quite critical of Merton's sociology of science. The chapter shows that Merton's sociological and Polanyi's epistemological approaches are coherent and mutually reinforce one another: their liberal defense of academic freedom and rejection of totalitarianism revolve around the idea that the values and the ethos of science should be respected as fundamental values in any liberal and democratic society – otherwise scientific research will be paralyzed and free society will collapse. Despite the short-comings of their positions, we should heed their warnings about the dangers of economic and political interference in science, especially as we are faced with similar, albeit less radical, attacks on the autonomy of science today.

The chapter is structured as follows. In Section 3.2, I analyze the intellectual relationship between Merton and Polanyi and show how they saw each other's views. While Merton sympathized with several of Polanyi's ideas, Polanyi had quite negative views of Merton's sociology of science, which he considered to be an erroneous relativistic theory. However, I argue that Polanyi misunderstood Merton's sociology of science. In fact, their views on the liberal values of science and their defense of academic freedom against totalitarian control are very similar. Both reject an anti-liberal conception of the state, according to which the government is solely responsible for the welfare of citizens, and thus the notion that publicly funded scientific research should be controlled and directed by a centralized government authority.

In Section 3.3, I show how Merton's famous conception of the values of the scientific community (*universalism, communalism, disinterestedness,* and *organized skepticism*) served as the basis for his defense of academic freedom against totalitarianism. I demonstrate that for Merton, the values of science are identical to the values of liberal and democratic societies. Much like Polanyi's, Merton's defense of academic freedom is essentially liberal. However, I also argue that Merton's position concerning external social and political influences on science was more historically adequate and empirically grounded than Polanyi's. Merton was aware of such external factors in scientific discoveries and did not share Polanyi's rigid distinction between applied and pure science. However, some critics have argued that Merton's conception of values is not helpful for understanding the operation of the scientific community. To respond, I contend that Merton's ideas can indeed be reformulated in a fruitful manner by relying on Radder's interpretation. I also show that Merton's account of academic freedom and liberal society is still relevant for understanding the political motivations behind the hostility toward scientific authority in the twenty-first century.

In Section 3.4, I reconstruct Polanyi's criticism of totalitarian control and his defense of academic freedom. I outline the historical context of Polanyi's position and analyze his objections to the idea of state-controlled or centrally planned science, as proposed by Marxist and socialist ideologists such as Bukharin and Bernal in the 1930s and 1940s. According to Polanyi, there are two main assumptions underlying totalitarian or authoritarian state control of science: scientific inquiry must be of immediate practical and economic value, and the state is solely responsible for the welfare of citizens. Polanyi argues that such assumptions inevitably lead to a rejection of the intrinsic value of scientific knowledge and to the notion that scientific research is valuable only if it serves the goals determined by a central authority. I present Polanyi's arguments for why he thinks these ideas are untenable and would paralyze scientific progress, and his claim that centralized state control will eventually not only lead to the collapse of academic freedom but also to that of free society in general. Polanyi argued that the state should recognize the autonomy of scientists and leave them to work on their problems, given that science can make progress only in spontaneous, unpredictable steps through autonomous interactions and information sharing between individual researchers.

Section 3.5 analyzes Polanyi's idea of a free society, namely his conception of public liberty as opposed to libertarian ideas, and reflects on some further objections to his account of academic freedom. I highlight that Polanyi's original account of academic freedom should be interpreted in its historical context. He developed his position as a response to the threat of twentieth-century totalitarianism, which is why some of his views may seem idealistic and elitist today. In particular, his sharp distinction between pure and applied science should be modified. Moreover, Polanyi's account lacks an analysis of ethical questions related to scientific research. Despite these problematic points, however, I argue that Polanyi's warnings about the political control of science are still relevant in democratic societies today. The motives behind state control of scientific research identified by both Merton and Polanyi continue to be present – most importantly, the view that scientific research must have immediate social or economic utility, and that the government is the sole or principal representative of the people's interests. These motives are independent of political systems and threaten academic freedom even in the twenty-first century and in countries where leaders are democratically elected.

3.2 Merton and Polanyi: Their Intellectual Relationship and Parallels in their Views

A comparison of Robert Merton's and Michael Polanyi's views on academic freedom and their criticism of totalitarian control of science would not seem to be straightforward if we considered only their intellectual relationship. Polanyi held quite negative views of Merton's approach to the sociology of science. Merton sent a copy of one of his papers to Polanyi ("Insiders and Outsiders: A Chapter in the Sociology of Science"), and after receiving it, Polanyi noted on the title page, somewhat sarcastically, that Merton was his "chief rival".[1] In fact, he also criticized Merton publicly. In his essay "On the Modern Mind," in a long footnote (Polanyi 1965, 18, fn. 4) on Harry M. Johnson's textbook entitled *Sociology: A Systematic Introduction*, Polanyi quoted Merton's introduction to the book and referred to him as a representative of the unacceptable value-free approach in sociology. Merton, on the other hand, approvingly quoted Polanyi's view on intellectual autonomy and free society in his essay "The Perspectives of Insiders and Outsiders" (Merton 1973a, 100, 136). In another essay, "Institutionalized Patterns of Evaluations in Science" (co-authored with Harriet Zuckerman), Merton also cited and approved of Polanyi's view of how scientific authority is established and functions based on a referee system (Merton 1973b, 461, 477, 491 fn. 53, 494). In addition, Merton referred to and cited Polanyi's main works, including *Personal Knowledge; Science, Faith, and Society; The Study of Man*; and *The Tacit Dimension*.

However, it is unlikely that Polanyi thoroughly read any of Merton's works. In contrast to Polanyi's characterization, Merton, like Polanyi, criticizes the relativistic tendencies in Mannheim's sociology of knowledge and links epistemic and moral relativism to totalitarian ideologies.[2] Polanyi's criticism of Merton does not account for this internal debate in the sociology of science between Merton and other key figures such as Mannheim and Durkheim.[3] Merton's essay, "Paradigm for the Sociology of Knowledge" (originally published in 1945 as "Sociology of Knowledge") is devoted to a critical engagement with Mannheim and the relativistic tendencies in the sociology of knowledge (Merton 1973d). Similarly to Polanyi, Merton identifies Mannheim's sociology as linked to the Marxist tradition. And like Polanyi, Merton also does not endorse its relativistic tendencies (Merton 1973d, 16–17, 21–22).[4] Furthermore, Merton interprets Durkheim's sociology as suggesting that the "criteria of validity" are determined by means of subjective and contingent sociological factors, meaning that for Durkheim, "Objectivity is itself viewed as a social emergent" (Merton 1973d, 25). Merton criticizes such a sociological approach as relying on a "dubious epistemology" (Merton 1973d, 25). Moreover, in "Sorokin's Formulations in the Sociology of Science" (co-authored with Bernard Barber), Merton also criticizes the relativistic tendencies of Sorokin's sociology (Merton 1973e, 163–166). All in all, as opposed to theories of the sociology of science and the strong program of the sociology of science, Merton downplays the external political, social, and ideological explanations for the acceptance or the rejection of scientific theories (Merton 1973f, 270).

My thesis is that despite Polanyi's negative view of Merton's sociology of science, their opposition to the totalitarian planning of science is actually very similar. Both defend the autonomy of pure science and free inquiry, reject central planning of scientific research, dismiss relativism in the sociology of science, and identify an essential connection between free society and the ethos of science (i.e., the moral and social norms and the values of the scientific community). Both recognize that the main assumptions of the totalitarian way of thinking about science arise from a collectivist view of the state and a vulgar form of utilitarianism. These assumptions can be summarized as follows. First, the state is solely responsible for the welfare of its citizens, and second, the only aim of science (along with other institutions) is to serve the practical and social needs determined by the state, which typically means the ruling party or the leader.

Both authors recognize that according to totalitarian or authoritarian ideology, the government is mainly, if not exclusively, responsible for the welfare of society, and that external political coordination and central planning of scientific research are therefore necessary. According to Polanyi's liberal vision, this conception of the state is a modern phenomenon and inevitably leads to the collapse of freedom in society (Mullins 2013, 163–165). One of the most infamous examples of totalitarian abuse of science is Lysenkoism in the Soviet Union (for a detailed analysis, see Polanyi 1951a, 59–65). In Polanyi's analysis, Lysenkoism was motivated by the rejection of the idea of pure science, i.e., the notion that scientific inquiries are valuable for the sake of knowledge itself even if they lack (immediate) practical or economic utility. Once this standpoint is accepted, it is only natural to take one step further and claim that all kinds of scientific inquiry must serve practical, economic interests (Polanyi 1951b, 3–4).

In his two seminal papers on the values of science and their relation to free society ("Science and the Social Order" 1938; "Science and Technology in a Democratic Order" 1942, reprinted as "The Normative Structure of Science") Merton, like Polanyi, reflects on the debates on the central planning of science initiated by Marxists-socialists in the 1930s. Analyzing the hostility against the autonomy of science in totalitarian states, Merton argues that conflict between such anti-liberal regimes and the values of the scientific community (most importantly, universalism, disinterestedness, communalism, and organized skepticism) is inevitable (Merton 1973f, 270–278). His examples of abuses of scientific norms are taken mainly from Nazi Germany, which systematically discriminated against so-called non-Aryan physicists (Merton 1973c, 255–257).

Before I analyze Merton's and Polanyi's arguments in more detail, let me quote from both authors to illustrate the basic congeniality of their criticism of planned science and government control:

> One sentiment which is assimilated by the scientist from the very outset of his training pertains to the purity of science. Science must not suffer itself to become the handmaiden of theology or economy or state. The function of this sentiment is to preserve the autonomy of science. For if such extrascientific criteria of the value of science as presumable consonance with religious doctrines or economic utility or political appropriateness are adopted, science becomes acceptable only insofar as it meets these criteria. In other words, as the pure science sentiment is eliminated, science becomes subject to the direct control of other institutional agencies and its place in society becomes increasingly uncertain. The persistent repudiation by scientists of the application of utilitarian norms to their work has as its chief function the avoidance of this danger, which is particularly marked at the present time.
>
> (Merton 1973c, 260)

> The new radically utilitarian valuation of science rests on a consistent philosophical background, borrowed mainly from Marxism. It denies that pure science, as distinct from applied or technical science, can exist at all. Such a revaluation of science necessarily leads to a demand for the Planning of Science. If science is to serve the practical needs of society it must be properly organized for this purpose. You cannot expect individual scientists, each pursuing his particular interests, to develop science effectively towards the satisfaction of existing practical needs. You must see to it therefore that scientists are placed under the guidance of authorities who know the needs of society and are generally responsible for safe-guarding the public interest.
>
> (Polanyi 1951c, 69)

As I will show, Merton and Polanyi defend the idea that scientists as free individuals must independently judge and make decisions on scientific matters (including scientific merit, validity, the plausibility of theories, and academic appointments), albeit without separating themselves from the rest of society. Both argue that scientists should submit themselves to norms and values that are driven by internalized obligations incorporated in the ethos of science, rather than being imposed on them by external authorities.

3.3 Merton on the Values and Autonomy of Science

From the 1930s and 1940s onward, Merton analyzed the social and philosophical sources of anti-intellectualism and the causes of the hostility toward science in totalitarian regimes. Apart from a general utilitarian approach to the role of scientific knowledge and the rejection of the value of pure science, Merton identified additional causes of anti-scientific measures. In his important paper, "Science and the Social Order" (1938), Merton shows that such anti-science policies (especially in Nazi Germany) originated from an inevitable conflict between the ethos of science and the aims of authoritarian political institutions (Merton 1973c, 257–260). Merton argues that abuses of the scientific community were initially an "unintended by-product of changes in political structure and nationalistic credo" (Merton 1973c, 255). That being said, scientists who did not meet the criteria of so-called "racial purity" or who were declared to be enemies for other political reasons suffered serious discrimination and were removed from universities and research institutions (Merton 1973c, 255).

Hostility toward scientific autonomy arises from the fact that scientific values, methods, and results are in conflict with the aims of non-liberal political institutions. Another reason why totalitarian regimes view academic freedom as a threat is that such regimes attempt to dispossess the public authority of science (Merton 1973f, 277), with state authorities mostly or solely in charge of the interpretation (or, misinterpretation) of scientific facts, rather than any autonomous scientific community (Merton 1973c, 264–266).

Merton's norms of scientific communities are part of his general account of liberal values in society and his criticism of authoritarian and totalitarian control of science. He claims that free and democratic societies accept, internalize, and protect the values of science, namely, *universalism, disinterestedness, communalism*, and *organized skepticism*. Universalism means that the acceptance or rejection of scientific claims should rely on universally binding norms instead of any personal or social attributes of their protagonists. Disinterestedness refers to the disinterested, impartial, and altruistic motives of scientific knowledge to the benefit to society. Communalism (or "communism", as it is sometimes called) refers to the communal ownership of scientific knowledge, meaning that intellectual property rights in science should be very limited. Organized skepticism refers to a critical attitude in science, namely that theories can and should be tested and that questioning and correcting each other's work is necessary and appreciated among scientists (Merton 1973f, 270–278).

Let us now turn to Merton's thesis that there is a strong connection between democratic principles and the values of science.[5] Merton identifies "the ethos of democracy" as one which "includes universalism as a dominant guiding principle" (Merton 1973f, 273).[6] According to *universalism*, scientific truth and the theoretical value of scientific claims are not dependent on race, gender, nationality, and any religious or political views. The acceptance or rejection of scientific statements must depend only on evidence. Analogously, in democracies, normative policy decisions are the result of deliberations in which the participants ideally evaluate each other's opinions based on evidence and rational arguments. As Kalleberg reminds us, deliberation, democracy, and the Mertonian norm of universalism are necessarily connected, since "normative claims can be cognitively valid, that is, defended, criticized, and decided with reasons" (Kalleberg 2010, 193).[7]

Totalitarian ideologies, however, discredited and repudiated scientific achievements on national, racial, or religious grounds. The totalitarian Nazi state put scientists under pressure to reject epistemic scientific standards of evaluating the plausibility or theoretical value of scientific claims. Scientists and scientific theories were labeled as "non-Aryan" or else (Merton 1973c, 260 fn. 20). By extension, universalism demands that the scientific community must be inclusive and open to all competent individuals regardless of their race, gender, ethnicity, or religion. Analogously, in free and democratic societies, all adult citizens are members of the *demos* and considered as free and equal (Kalleberg 2010, 194).

Although Merton's discussion of the epistemological underpinnings of democracy remained underdeveloped, it is not difficult to see why the other Mertonian values of science conflict with totalitarian politics. *Disinterestedness*, for example, requires individual scientists to consider arguments and to reject or revise their beliefs (Kalleberg 2010, 195). *Organized skepticism* demands "unrestricted discussions" (Kalleberg 2010, 195), which inevitably clashes with the policies imposed by authoritarian states, in which the decisions and creeds of political leaders are usually, if not always, incontestable and are not subject to evaluation by rational arguments and evidence. As Merton explains,

> Science, which asks questions of fact concerning every phase of nature and society, comes into psychological, not logical, conflict with other attitudes toward these same data which have been crystallized and frequently ritualized by other institutions. Most institutions demand unqualified faith; but the institution of science makes skepticism a virtue. ... It must be emphasized again that there is no logical necessity for a conflict between skepticism within the sphere of science and the emotional adherences demanded by other institutions. But as a psychological derivative, this conflict invariably appears whenever science extends its research to new fields toward

which there are institutionalized attitudes or whenever other institutions extend their area of control.

(Merton 1973c, 264–265)

The general lesson is that the internalized obligations and sentiments exemplified by the ethos of science are routinely abused by obligations and sentiments that the authoritarian or totalitarian state externally imposes on the scientific community. As Turner (2007, 173–175) has pointed out, Merton's criticism of government control and central planning of science can also be reconstructed in the following manner: external political control is unnecessary and problematic, because the internal scientific ethos of researchers guides them, in any case, to decide properly on scientific questions. (By contrast, the proponents of planned and centrally controlled science argue that all scientific activity should serve the goals imposed by state authorities.) Such values of science consist not only of intellectual but also of emotional components that sustain the norm-following behavior of the whole scientific community:

> This ethos, as social codes generally, is sustained by the sentiments of those to whom it applies. Transgression is curbed by internalized prohibitions and by disapproving emotional reactions which are mobilized by the supporters of the ethos. Once given an effective ethos of this type, resentment, scorn, and other attitudes of antipathy operate almost automatically to stabilize the existing structure.
>
> (Merton 1973c, 258 fn. 15)

According to Merton, the community of scientists and their internal norms and values should be protected from deleterious external (political, economic, or ideological) forces. Scientists should be committed to the general as well as to specific methodological norms. As Merton argues, these norms are "binding, not only because they are procedurally efficient, but because they are believed right and good" (Merton 1973f, 270).

Merton emphasizes the communal nature of scientific research, and he borrows the term "communism"/"communalism" in "The Normative Structure of Science" from Bernal (Turner 2007, 173–174). It should be noted that Merton quotes Bernal affirmatively in a few places and that his overall opinion of Bernal was not nearly as negative as Polanyi's.[8] However, this does not mean at all that Merton and Bernal share the same views on central planning and socialist collectivism. In his explication of the value of communalism, Merton focuses on Bernal's claim that "the growth of modern science coincided with a definite rejection of the ideal of secrecy" (Merton 1973f, 274 fn. 13). Turner offers a plausible reason why Merton emphasizes the social aspects of scientific activity and, especially, why scientific discoveries should not be the private intellectual property of individual scientists – namely because Merton's argument was liberal, but was presented "in the rhetorical clothing of the Left" (Turner 2007, 175). Similar to Polanyi, Merton's defense of the autonomy of science is essentially liberal (Turner 2007, 173). Like Polanyi, Merton emphasizes that scientific research and self-governance cannot be centralized and should remain autonomous. The following quote from "Science and the Social Order" (1938) illustrates this point:

> In the totalitarian society, the centralization of institutional control is the major source of opposition to science; in other structures, the extension of scientific research is of greater importance. Dictatorship organizes, centralizes, and hence intensifies sources of revolt against science that in a liberal structure remain unorganized, diffuse, and often latent.
>
> In a liberal society, integration derives primarily from the body of cultural norms toward which human activity is oriented. In a dictatorial structure, integration is effected primarily by formal organization and centralization of social control. ... The differences in the mechanisms through which integration is typically effected permit a greater latitude for self-determination and autonomy to various institutions, including science, in the liberal than in the totalitarian structure. Through such rigorous organization, the dictatorial state so intensifies its control over nonpolitical institutions as to lead to a situation that is different in kind as well as degree. ... In liberal structures the absence of such centralization permits the necessary degree of insulation by guaranteeing to each sphere restricted rights of autonomy and thus enables the gradual integration of temporary inconsistent elements.
>
> (Merton 1973c, 265–266)

Furthermore, science as a communal activity depends on expert opinions. Both Merton and Polanyi emphasize the importance of overlapping areas of experts. As will become clear in the following section, this is an important point in Polanyi's explanation of why central planning of science is practically impossible. As noted, Merton approvingly quotes Polanyi's view on the epistemic and social relevance of authority in science (Merton 1973b, 461). Merton shares Polanyi's view on the mutual aspect of scientific authority, namely that authority does not lie above scientists but "is established *between* scientists" (Polanyi 1962, 60). This supports the view that rather than any political platforms or doctrinal religious formulations, it is networks of experts in overlapping areas that establish the authority and stability of science.[9] Unlike Polanyi, however, Merton is aware that scientists, when seeking solutions to scientific problems, are sometimes motivated by social and economic factors even in pure sciences such as Newton's mechanics. Merton does not share Polanyi's sharp and radical distinction between pure and applied science. In his essay "Science and Economy of 17th Century England," Merton persuasively argues, for example, that the economic and military needs of navigation initiated and influenced the direction of "pure" physical research even if scientists themselves were not always aware of those external drivers. In addition, he also acknowledges the influence of other social factors on early modern science, such as military or economic considerations (Merton 1968, 662–663). As is well known, Merton, along the lines of Max Weber's classic studies on Protestant ethics and capitalism, reconstructed the social and historical connections between Protestant Puritanism and the scientific revolution in his early work *Science, Technology and Society in Seventeenth Century England* (Merton 1938), as well as in his later essays based on this study (for instance, "The Puritan Spur to Science," Merton 1973h). Nevertheless, this does not mean that Merton adopted the extreme Marxist view of the priority of applied science. For Merton, recognizing social and political factors and motivations does not entail the view that scientific claims should be assessed by their practical utility rather than by professional epistemic standards. This is because, as Merton contends, the "institutional goal of science is the extension of certified knowledge" from which institutional norms are derived (Merton 1973f, 270). Although not explicitly codified, these communal and institutional norms (relying on both epistemic and moral elements) are internalized as social and institutional norms, which are referred to as "the ethos of science" (Merton 1973f, 268–269).

Therefore, in line with Merton's account, a subtle distinction can be drawn between two types of external factors in scientific research: direct/ illegitimate and indirect/legitimate. Discriminating against scientists or scientific views on racial grounds, for example, is an obvious example of illegitimate external influence. However, some forms of influence are inevitable and not necessarily harmful (especially in practical, application-oriented inquiries). External influence on research should be deemed legitimate if scientists are motivated solely by scientific problems and are either not fully aware of or do not pay sufficient attention to external social influences. This crucial distinction between direct and indirect practical motivation also clearly distinguishes Merton's account from reductionist and socially deterministic accounts.[10]

In the past few decades, some have questioned whether Mertonian norms are indeed useful for understanding science as a social institution. Some critics (Barnes and Dolby 1970) have argued that these norms are not peculiar to science but are also observable in everyday life. Others, such as Mitroff (1974), have questioned to what extent the Mertonian norms are actually prescriptive for scientific communities. Sklair (1970) limits their scope to pure science and identifies different norms for applied science. Moreover, one might argue that Merton's analysis offers no detailed explanation of how these norms affect the cognitive goals of science (that is, the production of knowledge). As Stehr (1978, 185) argues, it is unclear whether the effect of social norms is direct or indirect. In any case, we need a functional explanation of the connection between the social and cognitive aspects of science.

However, more recently, some scholars have argued that Mertonian values are not outdated and continue to be useful for normative guidance of scientific research. To defend Merton against his critics, Radder (2010) distinguishes values from norms. Values are more universal and more vaguely defined than explicit norms. In Radder's conception, Merton's four primary ideas are values rather than strictly defined rules. For Radder, the role of general values is to give orientation to scientists by postulating the ends of scientific inquiry without providing a detailed explanation of how they should accept or reject different hypotheses. Specific professional and epistemic standards need to be fleshed out, and these specific rules tell scientists how to realize the general aims postulated by Mertonian values. Consequently, Radder (2010, 242–244) argues that Merton's values are important for several reasons. First, they strengthen the sense of community among scientists and thus deepen and invigorate public trust in science. Second, Mertonian values, which are unspecified and interdisciplinary, can be used as heuristic resources for the articulation of specific scientific norms. Third, even if Mertonian values do not determine scientists' inquiries in every detail, they guide them in their professional decisions.[11]

Furthermore, Merton's analysis of the causes of hostility toward academic freedom in Nazi Germany offers some important insights into present-day anti-science tendencies in some seemingly well-established liberal democracies. By applying a Mertonian analysis to these cases, it becomes clear that totalitarian attacks on academic freedom are based on two ideological assumptions. First, independent scientific results are seen to pose a danger to the realization of the state's aims as determined by the ruling party and the leader. Second, the ethos of the scientific community is seen as incompatible with the values or aims of the ruling party and the leader (Merton 1973c, 255). However, it is worth noting that such reasoning and the anti-science ideology it produces do not depend on whether a political leader is elected or not. Populist leaders, who are by definition democratically elected, typically equate the interests of the state with the so-called will of the majority and may thus view some scientific methods and results as antithetical to the interests of the society or the nation.

In sum, Merton's concept of democracy is more egalitarian and rooted in the American liberal tradition than Polanyi's conservative liberalism, which is inspired by his Hungarian, German, and British background. Still, as I will show in the next section, Polanyi's philosophically more detailed account of the values of a free society and the social and moral values of scientific communities shares several core points with Merton's liberal defense of academic freedom.

3.4 Polanyi's Case Against Totalitarianism and the Central Planning of Science

Polanyi's main ideas concerning the nature of scientific knowledge and free society, as well as his criticism of totalitarianism and central planning, were consistent throughout his life.[12] Polanyi's critical reflections on totalitarianism and his defense of academic freedom were a response to the socialist-Marxist conceptions of science and society developed in the 1930s and 1940s, mainly by John Desmond Bernal and Nikolai Bukharin. Socialist and Marxist authors criticized the idea of "pure science", i.e., the view that scientific knowledge is intrinsically valuable and that its value is thus independent of the promotion of immediate economic and practical benefits. Bernal, for example, claimed that such a conception of pure science is a form of "snobbery",[13] arguing that only applied science is valuable. Polanyi referred to his conversation about pure science with the Soviet Marxist Bukharin in 1935 as a decisive episode in his intellectual life (Polanyi 2017a, 63). Bukharin claimed that the distinction between pure and applied science assumed in capitalist societies was no more than an ideological construction that "deprived scientists of the consciousness of their social functions, thus creating in them the illusion of pure science" (Polanyi 2017a, 63).[14]

Following such Soviet tendencies of thought, the concept of planned science, where the government exerts extensive control over scientific research, and science is subordinated to the state's declared social aims, became increasingly popular in the United Kingdom in the 1930s. In 1931, an entire volume was dedicated to the Marxist understanding of so-called planned science, and British authors published a number of popular works on science from a Marxist or semi-Marxist point of view.[15] These books intensified the scholarly discussion of the social role of science in Britain (Turner 2007, 162–163). In 1938, the British Association for the Advancement of Science established a new division aimed at providing social guidance for the progress of science (Polanyi 1951b, 3 fn. 1). Additionally, the Great Depression of the 1930s provided another powerful reason for socialists to criticize both free-market capitalism and academic autonomy as understood in Western countries.

Polanyi's main thesis is that centralized control cannot work in practice, as it is based on an erroneous notion of scientific knowledge and discovery. In opposition to the totalitarian view of science and the British science planning movement, Polanyi argues that scientific research is a creative act whose outcome is unpredictable because it is a product of autonomous cooperation among researchers. Moreover, he argues that if the state takes control and science becomes an instrument of political-ideological goals, then such central planning (as in the economy) would inevitably paralyze scientific research regardless of how intelligent or benevolent the central authorities might be. Polanyi also states that a collapse of academic freedom leads to a collapse of freedom in society at large: "Academic freedom is of course never an isolated phenomenon. It can exist only in a free society; for the principles underlying it are the same on which the most essential liberties of society as a whole are founded" (Polanyi 1951d, 45).

In sum, Polanyi argues that the ideology of totalitarian control of science relies on two main assumptions: that every scientific inquiry must have immediate practical economic value, and that the state is solely responsible for the welfare of citizens. These assumptions entail a rejection of the intrinsic, non-instrumental value of scientific knowledge in all domains and the control of individual scientists by central directives of the government.

Polanyi argues that totalitarianism, by rejecting academic freedom, hinders us from pursuing scientific truth, not least because it denies the intrinsic value of the latter (as science ought to serve political or economic interests instead). Sooner or later, totalitarian control of science leads to a rejection of the assumption that scientific standards are the most reliable path toward truth, or even to a refutation of the existence of objective truth itself. As Polanyi argues,

> For if truth is not real and absolute, then it may seem proper that public authorities should decide what should be called the truth. ... [I]f our conceptions of truth and justice are determined by interests of some kind or another, then it is right that the public interest should overrule all personal interests in this matter. We have here a full justification of totalitarian statehood.
>
> (Polanyi 1951d, 47)[16]

If totalitarianism takes this final step and eliminates the intrinsic value of scientific truth, truth eventually becomes either unimportant or is merely identified as whatever satisfies political interests. To illustrate this point, Polanyi quotes Himmler's telling words on German pre-history:

> We don't care a hoot whether this or something else was the real truth about the pre-history of the German tribes. ... [T]here's no earthly reason why the party should not lay down a particular hypothesis as the starting point, even if it runs counter to current scientific opinion. The one and only thing that matters to us, and the thing these people are paid for by the State, is to have ideas of history that strengthen our people in their necessary national pride.
>
> (Quoted in Polanyi 1951a, 59)

Polanyi argues that such an anti-scientific attitude and practice is self-destructive both from a philosophical and a practical point of view. If one believes that there is no objective truth or that truth is solely contingent on its social and political usefulness, then it must be admitted that the ruling ideology itself is not objectively true either. Identifying truth with particular interests undermines the totalitarian ideology's claim of the falsity of its rival ideologies. If one accepts that a certain ideology is true, and yet argues that every ideology is a representation of special group interests, there is no reason to think that said ideology is an adequate description of reality (see Polanyi 1951/1952, 90–91). In practice, such a radical and ultimate form of totalitarianism will eventually collapse, because people cannot live in a prolonged state of complete intellectual schizophrenia and nihilism.[17]

Besides this general philosophical criticism of totalitarianism, Polanyi presents several objections against the central planning of science in particular, three of which I will reconstruct and highlight in the following.[18] First, scientific progress is unpredictable and relies on the autonomous and spontaneous cooperation of individual researchers. Second, tacit knowledge is indispensable in scientific knowledge. Third, scientific expertise is mutual, and the authority of science is therefore established by mutual relations and continuous interaction among scientists.

Polanyi's first argument against central planning relies on the core notion of *spontaneous* or *dynamic orders*. As Mullins (2013, 169) notes, Polanyi promotes a vision of "liberal society as largely a network of dynamic orders". Polanyi contrasts this *dynamic order* with *corporate (hierarchical) order* (Polanyi 1951e, 112, 114–115). He also refers to dynamic order as *spontaneous* and sometimes as *polycentric order* (Polanyi 1951f, 170–171).[19] Accordingly, Polanyi distinguishes between two kinds of coordination: *self-coordination (self-adjustment)* and *coordination by a central authority* (Polanyi 1951f, 170–171, 175–176). Self-coordination is a mutual adjustment of independent agents where every individual pays attention and adjusts to the operations of others within the same system (Polanyi 1941, 432–433, 441–443). Every agent acts freely, following his or her initiative, but in the sight of others and while responding to others' operations. Every single modification of the system takes into account all other modifications.[20] Similar to Hayek's criticism of the collectivist-socialist planning of the economy, Polanyi's criticism about the unpredictability of scientific research is based on his epistemological criticism of the collectivist view of knowledge production.[21] Moreover, Polanyi links his ideas to Adam Smith's theory of the economy and, by using a mathematical model, argues that many social tasks can be performed only by spontaneous coordination of free individuals.[22] In such systems, coordination should be polycentric rather than centralized.

There are simple and uniform tasks on which every individual works in the same manner. Polanyi's example is the task of shelling peas (Polanyi 1951d, 34). In such cases, individuals work separately and isolated from one another. However, scientific problem-solving is a different kind of task: complete isolation of the individuals would eliminate progress and prevent the solution of the problem. By contrast, Polanyi's example of a self-coordinating task is the solution of jigsaw puzzles (Polanyi 1951d, 35–36). Polanyi argues that even if a legion of puzzle-solvers were all working separately on different pieces of the puzzle, they would hardly be able to solve it effectively. And if they were subordinated into a hierarchic body, where a central authority directed their actions, the spontaneous cooperation between them would be paralyzed. Puzzle-solving and scientific inquiry are forms of coordination by self-adjustment: every time a piece is placed correctly, the other participants will consider the next step in light of the recent achievement. Solving a scientific problem is also a series of decisions where nobody knows or determines what the final outcome will be. Puzzle-solvers and scientists can only make incremental progress, and each consecutive step must be decided locally by competent individuals who keep a constant eye on the decisions of others (Polanyi 1951d, 34–35).

Polanyi also argues that the direction of scientific progress is unpredictable even for scientists themselves, because in a sufficiently large system, the number of possible actions an individual may take is enormous, so that a central authority is unable to control everything (Polanyi 1951f, 171–177). It is practically impossible, Polanyi contends, to give that many commands and to check whether the subordinated individuals are following them (Polanyi 1951e, 114–122). Goodman (2001, 11–12) reconstructs Polanyi's main idea as follows: the number of decisions needed for efficient coordination exceeds the cognitive capacity of any central planner to deal with them. In Goodman's reconstruction, Polanyi's other basic problem with central planning is that it requires too many decisions by a central authority, making it practically impossible to achieve efficient coordination.[23] Therefore, scientific problems cannot be solved in a hierarchical order coordinated by a central authority, but only if scientists are allowed to work on problems in a free, autonomous, and cooperative manner. Constraint by a centralized and hierarchical order would quickly paralyze the progress of scientific research (Polanyi 1951e, 119).

The second argument against central planning rests on Polanyi's conception of tacit knowledge (Tebble 2016, 7–8). In Polanyi's epistemology, tacit knowledge covers all non-propositional elements of knowledge. Such non-propositional knowledge cannot be explicitly articulated using explicit statements and cannot be explained by abstract logical or mathematical formulas. Skills and knowing-how are clear examples of tacit knowledge. Polanyi also includes intuitions, customs, creative acts, conjectures, hunches, and unspoken value commitments as important tacit elements that are indispensable for scientific discovery, the justification of scientific claims, and the evaluation of scientific merit (Polanyi 2005, 51–68, 142–158). These tacit elements of knowledge are necessarily connected to what Polanyi calls "intellectual passions", which motivate human beings to pursue scientific questions and engage in other intellectual activities (Polanyi 2005, 141–142). Polanyi argues that since scientific problem-solving tasks cannot be strictly formalized, even scientists themselves are at times unaware of the tacit elements of their knowledge. The progress of science depends on personal, tacit, and unpredictable judgments, making effective central planning practically impossible.[24]

The third argument against central planning can be summarized as follows: to control and govern scientific research efficiently, central authorities need to know the prospects of scientific inquiry in each discipline. However, no single individual, or even a small group of individuals, could possess the required skills and information because there are many overlapping research areas and the advancement of science is always piecemeal. Consequently, there is no general goal of scientific research taken as a whole (Polanyi 2017b, 132–133). Polanyi argues that the authority of science as an institution is established by individual experts in overlapping areas: every scientist is an expert only on a small fraction of scientific knowledge, with some secondary competence in certain proximate areas.[25]

Even assuming that there are competent members in central planning committees, they cannot possibly possess the information and skills needed for successful central coordination of scientific research, since it is practically impossible for the committee to have sufficient knowledge and understanding of all areas and topics of academic research.[26] This brings us to Polanyi's conception of the tacit elements of scientific discovery: "the methods of scientific inquiry cannot be explicitly formulated and hence can be transmitted only in the same way as an art, by the affiliation of apprentices to a master. The authority of science is essentially traditional" (Polanyi 1962, 69).[27] It is important to note that Polanyi's objection to central planning is *not only* that politicians are not qualified enough to evaluate scientific claims and are thus unable to effectively determine future research. His point is rather the following: central planning of scientific research *cannot* work because no single individual ever possesses the necessary relevant information, which is therefore never available to the planners, however intelligent they might be. Consequently, scientific progress will be paralyzed if scientists are guided by a central authority rather than freely sharing information and interacting with each other in solving scientific problems.

3.5 Polanyi's Vision of Free Society: Its Problems and Contemporary Relevance

As we have seen, Polanyi argues that the most efficient way to promote scientific progress is to let scientists work on their problems, by following their own, independent decisions and responding to each other's suggestions and previous achievements rather than to external directives. For Polanyi, the demand for central planning arises from two sources: first, the denial of the non-practical, intrinsic value of scientific knowledge, and second, an essentially anti-liberal conception of the state that is hostile to spontaneous order and the self-coordination of individuals (most importantly, in science).

It is not easy to separate Polanyi's vision of academic freedom and his conception of public liberty from his criticism of totalitarian central planning. Whereas Polanyi's analysis of the policies of totalitarian regimes and his warnings about the dangers of central, political control of science are typically not disputed, his conception of academic freedom has some shortcomings.

In this section, I will summarize how Polanyi's vision of free society depends on his overall account of academic freedom. In this reconstruction, I have three aims. First, to dissipate some misunderstandings of Polanyi's account by contrasting his conception of public liberty with an individualistic, libertarian conception of freedom. Second, to highlight some weak points in Polanyi's views, as a result of which his account of academic freedom and the relation between liberal society and the moral values of science should be modified and moderated, notwithstanding his forceful and valid criticism of totalitarianism. And finally, I also argue that similar to Merton's explanation of authoritarian regimes' hostility toward the ethos of science, Polanyi's warnings about the political control of science continue to be relevant because the main reasons underlying state control and the central planning of science are still present today, even in seemingly well-established liberal political systems. By appealing to the core ideas of Polanyi and Merton, it becomes possible to criticize the spread of anti-intellectualism, political voluntarism, and utilitarian conceptions of scientific research.

Polanyi's ideal of the scientific community as a free and autonomous society of individuals following transcendent values or ideals and producing spontaneous progress without a central authority is analogous to his vision of other social institutions and traditions such as free language, literature, arts, religion, British common law, or the free market (Polanyi 1941, 436–438, 448–449). Academic freedom and a free society depend on each other, and we cannot have one without the other. Therefore, as Polanyi states, "In the Liberal State the cultivation of science is [a] public concern" (Polanyi 2017b, 131). Polanyi's definition of academic freedom might be characterized simply as "the right to choose one's own problem of investigation, to conduct research free from any outside control, and to teach one's subject in the light of one's own opinions" (Polanyi 1951d, 33).

Polanyi argues that the government should recognize the value of pure science and should therefore exercise only "*supervisory* authority, presiding over the free individual initiatives" (Polanyi 1941, 439). He thinks that a liberal state must financially support scientific research but leave scientists to make their own decisions on research projects, appointments, and scholarship (Polanyi 1951d, 41–42). In Polanyi's conservative-liberal vision, the state operates like a good and wise monarch who provides economic support to scientists but does not tell them what to do.[28]

That being said, Polanyi does not endorse libertarian, state-free scientific research. His view of academic freedom should not be understood as a commitment to an almost unconstrained, individualistic, libertarian freedom. Instead of such individualistic conceptions of freedom (i.e., the freedom to pursue one's goals whatsoever), Polanyi defends the concept of *public liberty*: individual and institutional autonomy for the sake of pursuing ideals such as truth, justice, and beauty (Mullins 2013, 166–167).[29] Public liberty is "freedom with a responsible purpose; a privilege combined with duties" (Polanyi 1941, 438). Each scientist exercises public liberty by working on his/her own problems and contributing to the public discourse, which then produces the prevailing public opinion.

Scientists also have a responsibility to choose areas of investigation in line with their scientific ideas, but constrained by scientific values, traditions, and their membership of the "Society of Explorers" (Polanyi 1962, 72). Polanyi contends that such duties should not be imposed on scientists by an external authority, especially if that authority fails to respect these values. Although he does not reflect much on the problem of moral constraints of scientific research (such as morally problematic experiments on humans and animals), his conception does not exclude such constraints on academic freedom. The plausible reason for this is that for Polanyi, the overall value commitments of scientists to a free society involve fundamental moral commitments, equal freedoms, and human dignity.

Nevertheless, one problem still remains with Polanyi's theory. He presumes that scientists should be committed to liberal values for epistemic reasons. Without liberal values, there is no freedom of inquiry, and without freedom of inquiry, science cannot progress. However, the connection between the epistemic success of science and the underlying conservative-liberal values is looser than Polanyi thinks. For instance, a scientist who performs morally questionable medical experiments on humans could, in theory, produce novel, epistemically valuable scientific discoveries. Embryological or stem-cell research are other obvious examples of why we need to reflect on the moral constraints of scientific investigations, even in cases where they have great potential utility both epistemically and technologically. Polanyi's defense of academic freedom was a response to totalitarianism and he thus takes general liberal values for granted. Nonetheless, his account of academic freedom lacks a serious consideration of ethical problems, and he was perhaps too hasty in supposing that scientists are inevitably committed to liberal values as the only means to efficiently produce scientific knowledge.[30] Even if one agrees that totalitarian central planning of science should be rejected, one might argue that Polanyi's ideals were too idealistic even in his own time. Scholars frequently note that Polanyi's ideal of the scientific community (typically referred to as the "republic of science") was inspired by the Kaiser Wilhelm Institute where he served as a research fellow for many years (for historical details, see Nye 2011, 37–85). Polanyi depicts scientific institutions as a sort of spiritual community of individuals in which, like medieval monks, researchers dedicate themselves exclusively to the quest for truth while almost lacking any ulterior motivations (such as appreciation, fame, or money). However, scientific research (even in "pure science") is clearly motivated by scientists' desire and hope that their discoveries will someday be used for some practical purposes.

Another weak point in Polanyi's conception is the lack of historical support for his strict distinction between pure and applied science, and his denial that scientific discoveries are to some extent determined or influenced by social factors. Polanyi wrongly assumes a connection between two theses that are not necessarily connected. There is no strong link between the view that "science is a response to social needs" and the claim that "the validity of science ought to be judged by the degree of its serviceability to the ends of society" (Polanyi 1951–1952, 88). It does not logically follow that scientific claims cannot be justified on objective epistemic grounds *because* scientific research is influenced or motivated by the practical needs of society.

It is also worth noting that Polanyi's strict distinction between pure and applied science seems to be superfluous for his dialectical purposes. A successful criticism of central planning needs to take as a given that pure science is not *primarily* motivated by practical and economic needs and that practical utility in the fields of pure inquiry is *usually accidental* and cannot be clearly specified. To establish a more plausible account of the social role of science, we need to distinguish between direct, voluntary state intervention and indirect, collateral influence by the state. While Polanyi rightfully criticizes the former, the latter is inevitable and typically does not pose a danger to scientific freedom. Concerning this point, Merton's analysis of direct and indirect influence on science is more plausible, as well as being supported by sociological and historical evidence.

Nevertheless, despite the shortcomings of Polanyi's position, his objections to the anti-liberal tendencies of state control of science should remind us that authoritarianism can also be a product of an appeal to the will of a "democratic" majority. The same argument for state control of science and the subordination of scientific research to political goals may be raised in countries where leaders are democratically elected.[31]

The problem is that democratically elected government officials may interpret the principles of academic autonomy as a subversion of the so-called will of the people.[32] After all, the view that the state is (solely) responsible for the welfare of society and that state-funded scientific research must therefore follow government directives can be stated perhaps even more persuasively in democratic societies where leaders are elected by universal suffrage. It seems tempting to declare that the people's interests are best represented by a government elected by the majority of the people, and that state-funded scientific institutions should therefore be controlled and directed by the state, rather than by unelected researchers representing only their own interests. Here, it is worth recalling Polanyi's reconstruction of the main argument for the necessity of state control of science:

> If science is to serve the practical needs of society it must be properly organized for this purpose. You cannot expect individual scientists, each pursuing his particular interests, to develop science effectively towards the satisfaction of existing practical needs. You must see to it therefore that scientists are placed under the guidance of authorities who know the needs of society and are generally responsible for safe-guarding the public interest.
>
> (Polanyi 1951c, 69)

Various labels can be used for such political goals, such as "social progress", "the interest of the nation", or even "democracy". To avoid problems of totalitarian control in democracies, we should be able to distinguish between acceptable (or even desirable) and inadmissible cases of so-called "democratic control" of science. For example, it is widely accepted that academic appointments and scholarships should not be determined by popular referendum or by (elected) government officials.

It is worth recalling Polanyi's warning that a liberal society is a society that is already committed to the beliefs that are necessary for freedom, including the commitment to academic freedom. In other words, a liberal society should not be value-neutral on such fundamental matters (see Roberts 1969, 237). These ideas, Polanyi argues, should be considered as being "transcendental" to society. Whereas Polanyi is not against universal suffrage, he insists that democracy needs principles that can never be overruled even by a decision of the majority (Polanyi 2017a, 67–68). As Polanyi warns us, totalitarian and autocratic leaders typically promise that "the central planning of science, and of other cultural and economic activities, would not be oppressive, since it would be based on democratic elections with a wide franchise" (Polanyi 1941, 443). By contrast, Polanyi contends that the principle of democracy is not merely about elections, but "must be based on the proper division of the social order between the corporate and dynamic forms of organisation" (Polanyi 1941, 443).

In sum, Polanyi, similarly to Merton, observes that any regime (whether democratically elected or not) might declare that it is either practically or morally necessary (or at least admissible) to directly control scientific research and researchers. This serves as a reminder that even democratically elected leaders may have a political interest in attacking scientific authority and might therefore attempt to dispossess the authority of science, much like the totalitarian regimes of the twentieth century did before them.

3.6 Conclusions

This paper contextualized and compared Robert Merton's and Michael Polanyi's arguments in favor of academic freedom and against totalitarianism. I analyzed these arguments in the context of both authors' overall accounts of the appropriate social, moral, and epistemic values for the scientific community, often referred to as the ethos of science. In doing so, I identified Merton's values for science (*universalism, organized skepticism, communalism*, and *disinterestedness*) as essential elements of his broader, essentially liberal vision of both the scientific community and a free society as a whole. Although Merton's conception of the ethos of science does not involve a carefully articulated normative epistemology, his sociology of science is compatible with normative considerations of scientific justification and the value of scientific knowledge. And while Merton's sociology does not posit an epistemic concept along the lines of Michael Polanyi's "tacit knowledge", his account fits reasonably well with the latter's general epistemological ideas.

Although Merton's sociological and Polanyi's epistemological arguments complement one other, Polanyi himself viewed Merton's work negatively and worried that Merton embraced a purely descriptive, value-free sociology of knowledge. However, Polanyi's interpretation of Merton is questionable for many reasons. First, Merton himself was critical of the relativist tendencies in the sociology of knowledge, which Polanyi represented as instances of a descriptive sociology of science. And second, Polanyi's conception of moral and social norms that are binding for scientists in fact has strong parallels with Merton's central ideas.

Both Merton and Polanyi agree on the rejection of central and authoritarian planning of science. Although Merton's ideas about the connections between the values of science and the values of democracy are underdeveloped, he suggests (similarly to Polanyi) that the basic values establishing a free society are integral to the practices and commitments operating in the scientific community (most importantly, universalism and organized skepticism, to use Merton's terminology). Freedom collapses if academic autonomy is not guaranteed. For Merton, similarly to Polanyi, a free society should endorse the same values as science. The guiding principles of the scientific community are thus a model for democratic society as a whole.

On the other hand, I also pointed out a number of important differences between Merton's and Polanyi's views. Polanyi's approach is more philosophical than Merton's, and his account of academic freedom and thus of the ethos of science is centered more on epistemological and metaphysical questions. Consequently, his criticism of totalitarianism is integral to his broader philosophical account of modern science and his critique of basic modern ideas such as skepticism, positivism, and utilitarianism, which, according to Polanyi, fueled modern totalitarian ideologies. However, Polanyi's philosophical abstractions often lead him to be less sensitive to empirical-historical issues in history and the sociology of science. Most importantly, Merton's conception concerning the relation between pure and applied science is more down-to-earth and empirically grounded than Polanyi's (almost) Platonist vision. Unlike Polanyi, Merton recognizes the influence of social, economic, and religious factors on the progress of science.

Furthermore, I also reconstructed Polanyi's arguments against central planning and his criticism of totalitarianism based on his epistemology. Similarly to liberal critics of the collectivist economy (like Hayek), Polanyi argues that such guidance not only leads to the collapse of freedom but also does not work in practice. According to Polanyi, there is an essential connection between the ideas of totalitarianism, collectivist planning, the rejection of the value of pure science as merely a truth-seeking activity, and the erroneous presumption that scientific research activities should have immediate practical, economic, or even political benefits. Polanyi argues that totalitarian ideology denies the value of scientific truth and the authority and freedom of scientists to make judgments on scientific questions and choose their own problems guided by their own internal values and norms. In Polanyi's vision, a free society will respect and nurture academic freedom as perhaps the most important form of "public liberty". While the state should support and supervise scientific research, the progress of science is and should be guided by spontaneous self-coordination among individual scientists rather than by any central authority.

I showed that both Merton's and Polanyi's defense of the classical autonomy of science have provoked criticism. Some might argue that the complete, unconstrained autonomy of science is not a desirable goal even in democratic and free societies. Finally, despite the shortcomings of their positions, I defended the thesis that Merton's and Polanyi's warnings of state control of science are still relevant today. Demands for state control can be convincingly based on an appeal to democratic elections presented as "the will of the people". If politicians declare that all state-funded institutions (including scientific institutions) should serve the interests of the people or the nation, and that the elected government alone represents the people's interests, this could easily lead to the conclusion that government control of scientific research is necessary and legitimate.

Polanyi and Merton are right when they identify the foundation of such reasoning as based on an essentially anti-liberal conception of the state. The recent ascendancy of anti-intellectual movements, even in seemingly well-established liberal political systems, has shown that the rejection of the autonomous ethos of science and the endorsement of a radical utilitarian view of scientific knowledge are not limited to totalitarian states. Of course, positing an infallible and sacrosanct status for science is not a remedy against political populism or cynical utilitarian views of scientific knowledge. The lesson to be taken from Merton's and Polanyi's criticism of totalitarian control of science is that a free society should defend the basic commitment to truth as valuable and grounding, and that only free and autonomous researchers are able to effectively explore objective reality. While it should be presented more carefully, a general commitment to basic epistemic and non-epistemic values (the ethos of science) remains essential for a free society, and arguably for any practically functioning modern society. Even if Merton's and Polanyi's classic ideals of academic freedom should be modified in a way that is more suited to twenty-first-century circumstances, their defense of the ethos of science against state control could provide valuable lessons for present-day discussions about the anti-science bias in authoritarian politics.

Acknowledgments

This chapter contributes to the research program of the MTA BTK Lendület Morals and Science Research Group. My research was supported by the MTA BTK Lendület Morals and Science Research Group. The paper is an extended and amended version of a conference talk delivered at the *Science, Freedom, Democracy* conference at the Institute of Philosophy, Research Centre for the Humanities, Hungarian Academy of Sciences, in Budapest on July 8–9, 2019. The conference was supported by the Hungarian Academy of

Sciences – International Conference Fund and MTA BTK Lendület Morals and Science Research Project. I would like to express my special thanks to Phil Mullins, Ádám Tamás Tuboly, and Stephen Turner for their helpful comments and critical remarks.

Notes

1 I would like to thank to Phil Mullins for these points. I am more than grateful to him for his invaluable comments throughout this paper.

2 "It is of considerable interest that totalitarian theorists have adopted the radical relativistic doctrines of *Wissenssoziologie* as a political expedient for discrediting 'liberal', or 'bourgeois' or 'non-Aryan' science. … Politically effective variations of the 'relationism' of Karl Mannheim (for example, *Ideology and Utopia*) have been used for propagandistic purposes by such Nazi theorists as Walter Frank, Krieck, Rust, and Rosenberg" (Merton 1973c, 260 fn. 20).

3 Polanyi and Mannheim corresponded with one other. In a letter to Mannheim written on April 19, 1944, Polanyi criticized Mannheim's position as social determinism (see Jacobs and Mullins 2006, 150–151).

4 Without explicitly mentioning Mannheim, Polanyi sharply criticizes relativist theories of historical understanding and value-free assumptions of the humanities in his lecture on "Understanding History", published in *The Study of Man* (Polanyi 1959, 71–73, 78–85).

5 However, this thesis should not be confused with the view that scientific progress is impossible when the leaders of the state are unelected. Merton reminds us that past monarchs and hereditary aristocracies also supported science (Merton 1973f, 269).

6 Merton often uses the terms "democracy" and "liberalism" interchangeably. Given the elusiveness of the notion of democracy, from an analytical point of view, it would have been useful for him to carefully distinguish between free, liberal society and democracy. Most prominently, Tocqueville exposed the conceptual and practical tensions between majoritarian, democratic ideas and liberalism (for the historical changes in the notion of democracy and the problem of "democratic despotism", see Nyirkos [2018]). Nevertheless, analyzing the many different meanings of "democracy" would be beyond the scope of this chapter. Therefore, risking simplification, I adhere to Merton's somewhat loose terminology. In the following, by "democracy", I mean a constitutional liberal society with inclusive suffrage and elections.

7 My claim here is not that Merton had a developed account of democratic deliberation similar to Habermas, for example.

8 For instance, see Merton's analysis of the so-called gerontocracy in science in "Age, Aging, and Age Structure in Science" (Merton 1973g, 538).

9 "Although rank and authority are *acquired* through past performance, once acquired they then tend to be *ascribed* (for an indeterminate duration). This combination of acquired and ascribed status introduces strains in the operation of the authority structure of science, as has been noted with great clarity by Michael Polanyi and Norman Storer" (Merton 1973b, 477).

10 "It is neither an idle nor unguarded generalization that *every English scientist of this time* who was of sufficient distinction to merit mention in general histories of science at one point or another explicitly related at least some of his scientific research to immediate practical problems. But in any case, analysis exclusively in terms of (imputed) motives is seriously misleading and tends to befog the question of the modes of socio-economic influence upon science. It is important to

distinguish the personal attitudes of individual men of science from the social role played by their research. Clearly, some scientists were sufficiently enamored of their subject to pursue it for its own sake, at times with little consideration of its practical bearings. Nor need we to assume that *all* individual researchers are directly linked to technical tasks. The relation between science and social needs is two-fold: direct, in the sense that some research is advisedly and deliberately pursued for utilitarian purposes and indirect, in so far as certain problems and materials for their solution come to the attention of scientists although they need not be cognizant of the practical exigencies from which they derive" (Merton 1968, 662–663).

11 Radder (2010, 252) calls his defense of Merton deflationary and neo-Mertonian.

12 Apart from his magnum opus *Personal Knowledge* (first published in 1958), these interconnected ideas and arguments were presented in several essays, lectures, and collections of papers including *The Growth of Thought in Society* (1941), *Science, Faith and Society* (1946), *The Logic of Liberty* (1951), the *Gifford Lectures* (1951–1952), *The Republic of Science* (1962), as well as in several other essays collected in *Knowing and Being, Essays by Michael Polanyi* (edited by Marjorie Grene) and *Society, Economics, and Philosophy: Selected Papers* (edited by R.T. Allen).

13 Polanyi quotes Bernal in his "Rights and Duties of Science" (Polanyi 2017a, 69).

14 As another Marxist ideologist, Mitin, stated, "We have no gulf between theory and practice, we have no Chinese wall between scientific achievements and practical activity. Every genuine discovery, every genuine scientific achievement is with us translated into practice … Soviet biologists, geneticists and selectionists must understand dialectical and historical materialism, and learn to apply the dialectical method to their scientific work. Verbal, formal acceptance of dialectical materialism is not wanted" (Polanyi quotes Mitin in his "Self-Government of Science", Polanyi 1951a, 62).

15 To name but a few: Hogben's *Mathematics for the Million* (1937), Hyman Levy's *The Universe of Science* (1933) and *Modern Science: A Study of Physical in the World Today* (1939), and J.B.S. Haldane's *Science and Everyday Life* (1940).

16 As opposed to Polanyi, Neurath argues that commitments to objective and absolute truth can easily lead to a totalitarian way of thinking (for an analysis of the parallels and differences between Polanyi's and Neurath's philosophy and their life, see Tuboly [2020]). I would like to thank to Ádám Tamás Tuboly for this point.

17 Polanyi called the Hungarian Revolution in October 1956 a battle for truth itself. What caused the Hungarian Revolution in Polanyi's view was that the total and radical denial of truth had become unsustainable. People revolted not only for political freedom but for the principle that objective truth exists and is valuable. He quotes the Hungarian communist Miklós Gimes's words to illuminate the radical totalitarian subversion of truth and the subsequent transformation of this totalitarian viewpoint into the normal mentality: "Slowly we had come to believe … that there are two kinds of truth, that the truth of the Party and the people can be different and can be more important than the objective truth and that truth and political expediency are in fact identical … And so we arrived at the outlook … which poisoned our whole public life, penetrated the remotest corners of our thinking, obscured our vision, paralysed our critical faculties and finally rendered many of us incapable of simply sensing or apprehending truth. This is how it was, it is no use denying it" (Polanyi 1969, 29).

18 The reconstruction that follows relies mainly on studies by Goodman (2001), Hartl (2012), Mullins (2013), and Jacobs and Mullins (2008).

19 The concept of "spontaneous order" plays a key role in both Hayek and Polanyi (see Jacobs 2000). Their ideas are very similar, but it was Polanyi who used the

term first in his 1941 paper "The Growth of Thought in Society" (see Jacobs and Mullins 2008, 122 fn. 11; and also Tebble 2016, 22 fn. 31).

20 See also Polanyi (1951d, 34–36) and (1962, 54–57).

21 Polanyi's criticism of central planning in science is in many aspects similar to Hayek's criticism of collectivist planning in the economy. In Chapter 11 of *The Road to Serfdom*, "The End of Truth", Hayek's criticism of totalitarian control of science is very similar to Polanyi's (Hayek 2001, 157–171). Polanyi also wrote a favorable review of Hayek's *The Road to Serfdom* (*The Spectator*, 1944, 31 March, p. 293). As Mullins (2013, 163) informs us, Polanyi and Hayek became friends in the late 1930s and corresponded with each other for around 30 years (Jacobs and Mullins 2016). For a comparison of Hayek's and Polanyi's philosophy, see also Mirowski (1998).

22 "[T]he coordinating functions of the market are but a special case of coordination of mutual adjustment" (Polanyi 1962, 56). See also Polanyi (1951f, 170).

23 Goodman (2001, 12) distinguishes Polanyi's argument about polycentricity from Mises's and Hayek's criticism of socialist central planning of the economy.

24 "You can kill or mutilate the advance of science, you cannot shape it. For it can advance only by essentially unpredictable steps, pursuing problems of its own, and the practical benefits of these advances will be incidental and hence doubly unpredictable" (Polanyi 1962, 62).

25 "But the authority of scientific opinion remains essentially mutual; it is established *between* scientists, not above them. Scientists exercise their authority over each other" (Polanyi 1962, 60).

26 "Scientific opinion is an opinion not held by any single human mind, but one which, split into thousands of fragments, is held by a multitude of individuals, each of whom endorses the other's opinion at second hand, by relying on the consensual claims which link him to all the others through a sequence of overlapping neighbourhoods" (Polanyi 1962, 59–60).

27 Polanyi is often criticized for being elitist and holding potentially authoritarian epistemological assumptions. Another eminent Hungarian philosopher of science, Imre Lakatos, presents this criticism perhaps most forcefully. While Polanyi interprets, for example, the victory of Lysenko over the Mendelians as an abuse of the norms of the scientific community, Lakatos argues that such social norms of the community do not guarantee scientific progress, i.e., an ever more comprehensive understanding of nature. Consensus among scientists and degeneration in terms of epistemic success can prevail simultaneously (Lakatos 1978a, 111–117). Lakatos argues that Polanyi's view is not able to explain how scientific progress is possible if we have nothing else to rely on than scientists' inexplicable intuitive judgments. If the only judges of scientific truth are members of the privileged elite, then the progress of science is not rational, Lakatos maintains, and, therefore, such appeal to scientific authority could open the door to arbitrary decisions and a form of authoritarianism (Lakatos 1978a, 112–120; Lakatos 1978b, 227–228). It is worth noting that Lakatos also sharply criticizes Merton. According to Lakatos, Merton's sociological approach cannot give a satisfactory answer to the demarcation question, namely what distinguishes science from pseudoscience (see Lakatos 1978a, 114–116).

28 At least Polanyi's arguments for government support of the economy and subsidized science are not inconsistent: he doesn't defend laissez faire in either case, even though he more frequently criticizes it in the economic sphere than in the sphere of science. Nevertheless, Polanyi's analogy between the free market and scientific research is problematic because it overlooks some important factors. Whereas science aims at truth, a consumer society established by the free market often manipulates customers. Unlike science, the free market flourishes in part

because its advancement relies, to some extent, on manipulative advertising and aggregated demand (see also Hartl 2012, 318–319).

29 Polanyi repeatedly criticizes laissez faire liberalism by arguing that it causes unemployment. According to Polanyi, the state's failure to intervene made the Great Crisis even worse (Polanyi 2017b, 139–140). Polanyi's idiosyncratic economic theory can be characterized as a middle-of-the-road position between the Austrian School of Economics and Keynesianism. For a detailed analysis of Polanyi's economic views and his criticism of central planning, see Bíró (2019) and Bíró (forthcoming).

30 One might also object that in the Soviet Union, planned science was actually successful in some areas. Soviet science and engineering ranked highly and could compete with Western science in some disciplines. Polanyi himself addresses this objection, insisting that these successes were due to the fact that centralized control in these areas was not implemented in practice but remained on paper only (Polanyi 1951c, 83–85).

31 For an elaborate defense of a form of democratic control of science that is based both on respect of egalitarian-liberal principles and a recognition of the value of scientific truth, see Kitcher (2001).

32 "When we reject today the interference of political or religious authorities with the pursuit of science, we must do this in the name of the established scientific authority which safeguards the pursuit of science" (Polanyi 1962, 68).

References

Barnes, S.B., and R.G.A. Dolby. 1970. "The Scientific Ethos: A Deviant Viewpoint." *European Journal of Sociology* 11 (1): 3–25.

Bíró, G. 2019. *The Economic Thought of Michael Polanyi*. New York and London: Routledge.

Bíró, G. (forthcoming). "From Red Spirit to Underperforming Pyramids and Coercive Institutions: Michael Polanyi Against Economic Planning." *History of Political Economy*.

Goodman, C.P. 2001. "A Free Society: The Polanyian Defence." *Tradition and Discovery* 27 (2): 8–25.

Hartl, P. 2012. "Michael Polanyi on Freedom of Science." *Synthesis Philosophica* 27 (2): 307–321.

Hayek, F.A. 2001. *The Road to Serfdom*. London and New York: Routledge.

Jacobs, S. 2000. "Spontaneous Order: Michael Polanyi and Friedrich Hayek." *Critical Review of International Social and Political Philosophy* 3 (4): 49–67.

Jacobs, S., and P. Mullins. 2006. "T.S. Eliot's Idea of the Clerisy, and its Discussion by Karl Mannheim and Michael Polanyi in the Context of J. H. Oldham's Moot." *Journal of Classical Sociology* 6 (2): 147–156.

Jacobs, S., and P. Mullins. 2008. "Faith, Tradition, and Dynamic Order: Michael Polanyi's Liberal Thought from 1941 to 1951." *History of European Ideas* 34 (1): 120–131.

Jacobs, S., and P. Mullins. 2016. "Friedrich Hayek and Michael Polanyi in Correspondence." *History of European Ideas* 42 (1): 107–130.

Kalleberg, R. 2010. "The Ethos of Science and the Ethos of Democracy." In *Robert K. Merton. Sociology of Science and Sociology as Science*, edited by C. Calhoun, pp. 182–213. New York: Columbia University Press.

Kitcher, P. 2001. *Science, Truth, and Democracy*. Oxford: Oxford University Press.

Lakatos, I. 1978a. "The Problem of Appraising Scientific Theories: Three Approaches." In *Imre Lakatos: Mathematics, Science and Epistemology. Philosophical Papers*, vol. II, edited by J. Worall and G. Currie, pp. 107–120. Cambridge: Cambridge University Press.

Lakatos, I. 1978b. "Understanding Toulmin." In *Imre Lakatos: Mathematics, Science and Epistemology. Philosophical Papers*, vol. II, edited by J. Worall and G. Currie, pp. 224–243. Cambridge: Cambridge University Press.

Merton, R.K. 1938. "Science, Technology and Society in Seventeenth Century England." *Osiris* 4: 360–632.

Merton, R.K. 1968. "Science and Economy of 17th Century England." In *Social Theory and Social Structure*, pp. 661–681. New York: Free Press.

Merton, R.K. 1973a. "The Perspectives of Insiders and Outsiders." In *The Sociology of Science: Theoretical and Empirical Investigations*, edited and with an introduction by Norman W. Storer, pp. 99–136. Chicago: University of Chicago Press.

Merton, R.K. 1973b. "Institutionalized Patterns of Evaluations in Science" (co-authored with Harriet Zuckerman). In *The Sociology of Science: Theoretical and Empirical Investigations*, edited and with an introduction by Norman W. Storer, pp. 460–496. Chicago: University of Chicago Press.

Merton, R.K. 1973c. "Science and the Social Order." In *The Sociology of Science: Theoretical and Empirical Investigations*, edited and with an introduction by Norman W. Storer, pp. 254–266. Chicago: University of Chicago Press.

Merton, R.K. 1973d. "Paradigm for the Sociology of Knowledge." In *The Sociology of Science: Theoretical and Empirical Investigations*, edited and with an introduction by Norman W. Storer, pp. 7–40. Chicago: University of Chicago Press.

Merton, R.K. 1973e. "Sorokin's Formulations in the Sociology of Science" (co-authored with Bernard Barber). In *The Sociology of Science: Theoretical and Empirical Investigations*, edited and with an introduction by Norman W. Storer, pp. 142–172. Chicago: University of Chicago Press.

Merton, R.K. 1973f. "The Normative Structure of Science." In *The Sociology of Science: Theoretical and Empirical Investigations*, edited and with an introduction by Norman W. Storer, pp. 267–278. Chicago: University of Chicago Press.

Merton, R.K. 1973g. "Age, Aging, and Age Structure in Science" (co-authored with Harriet Zuckerman). In *The Sociology of Science: Theoretical and Empirical Investigations*, edited and with an introduction by Norman W. Storer, pp. 497–559. Chicago: University of Chicago Press.

Merton, R.K. 1973h. "The Puritan Spur to Science." In *The Sociology of Science: Theoretical and Empirical Investigations*, edited and with an introduction by Norman W. Storer, pp. 228–253. Chicago: University of Chicago Press.

Mirowski, P. 1998. "Economics, Science and Knowledge: Polanyi vs Hayek." *Tradition and Discovery* 25 (1): 29–42.

Mitroff, I. 1974. "Norms and Counter-Norms in a Select Group of the Apollo Moon Scientists: A Case Study of the Ambivalence of Scientists." *American Sociological Review* 39 (4): 579–595.

Mullins, P. 2013. "Michael Polanyi's Early Liberal Vision: Society as a Network of Dynamic Orders Reliant on Public Liberty." *Perspectives on Political Science* 42 (3): 162–171.

Nye, M.J. 2011. *Michael Polanyi and His Generation: Origins of the Social Construction of Science*. Chicago and London: University of Chicago Press.

Nyirkos, T. 2018. *The Tyranny of Majority: History, Concepts, and Challenges.* New York and London: Routledge.

Polanyi, M. 1941. "The Growth of Thought in Society." *Economica* 8 (32): 428–456.

Polanyi, M. 1951–1952. *Gifford Lectures* (edited and with an introduction by Phil Mullins). Available online at www.polanyisociety.org/Giffords/Giffords-web-pa ge9-20-16.htm. Accessed January 18, 2020.

Polanyi, M. 1951a. "Self-Government of Science." In *The Logic of Liberty: Reflections and Rejoinders*, pp. 49–67. London: Routledge and Kegan Paul.

Polanyi, M. 1951b. "Social Message of Pure Science." In *The Logic of Liberty: Reflections and Rejoinders*, pp. 3–7. London: Routledge and Kegan Paul.

Polanyi, M. 1951c. "Science and Welfare." In *The Logic of Liberty: Reflections and Rejoinders*, pp. 68–85. London: Routledge and Kegan Paul.

Polanyi, M. 1951d. "Foundations of Academic Freedom." In *The Logic of Liberty: Reflections and Rejoinders*, pp. 32–48. London: Routledge and Kegan Paul.

Polanyi, M. 1951e. "The Span of Central Direction." In *The Logic of Liberty: Reflections and Rejoinders*, pp. 111–137. London: Routledge and Kegan Paul.

Polanyi, M. 1951f. "Manageability of Social Tasks." In *The Logic of Liberty: Reflections and Rejoinders*, pp. 154–200. London: Routledge and Kegan Paul.

Polanyi, M. 1959. "Understanding History." In *The Study of Man*, pp. 71–99. Chicago: University of Chicago Press.

Polanyi, M. 1962. "The Republic of Science: Its Political and Economic Theory." *Minerva* 1: 54–73.

Polanyi, M. 1965. "On the Modern Mind." *Encounter* 24: 12–20.

Polanyi, M. 1969. "The Message of the Hungarian Revolution." In *Knowing and Being. Essays by Michael Polanyi*, edited by Marjorie Grene, pp. 24–39. Chicago: University of Chicago Press.

Polanyi, M. 2005. *Personal Knowledge: Towards a Post-critical Philosophy*. London: Routledge.

Polanyi, M. 2017a. "Rights and Duties of Science." In *Society, Economics, and Philosophy. Selected Papers: Michael Polanyi*, edited by R.T. Allen, pp. 61–77. London and New York: Routledge.

Polanyi, M. 2017b. "Collectivist Planning." *Society, Economics, and Philosophy. Selected Papers: Michael Polanyi*, edited by R.T. Allen, pp. 121–143. London and New York: Routledge.

Radder, H. 2010. "Mertonian Values, Scientific Norms, and the Commodification of Scientific Research." In *The Commodification of Scientific Research: Science and the Modern University*, edited by H. Radder, pp. 231–258. Pittsburgh: University of Pittsburgh Press.

Roberts, P.C. 1969. "Politics and Science: A Critique of Buchanan's Assessment of Polanyi." *Ethics* 79 (3): 235–241.

Sklair, L. 1970. "The Political Sociology of Science: A Critique of Current Orthodoxies." *The Sociological Review* 18 (1) Supplement: 43–59.

Stehr, N. 1978. "The Ethos of Science Revisited: Social and Cognitive norms." *Sociological Inquiry* 48 (3–4): 172–196.

Tebble, A.J. 2016. *Epistemic Liberalism. A Defence.* New York and London: Routledge.

Tuboly, A.T. 2020. "Michael Polanyi and Otto Neurath: An Unplanned Parallel in British Intellectual Life: Review of Gábor Bíró: *The Economic Thought of Michael Polanyi.*" *History of European Ideas* 46 (2): 218–224.

Turner, S. 2007. "Merton's 'Norms' in Political and Intellectual Context." *Journal of Classical Sociology* 7 (2): 161–178.

4 Scientific Freedom and Social Responsibility

Heather Douglas

DEPARTMENT OF PHILOSOPHY, MICHIGAN STATE UNIVERSITY

4.1 Introduction

Over the past two decades, we have witnessed a sea change in the understanding of scientific freedom and social responsibility among scientists. Scientists have gone from thinking of freedom and responsibility as in tension, such that more of one meant less of the other, to being yoked together, such that more of one meant more of the other. This is a crucial and important change, and one that is easy to lose sight of in the midst of debates over particular scientific practices. I will show in this chapter that such a change has occurred (drawing on the central documents and scientific institutions that reflect scientists' understanding of these issues) and articulate some reasons why such a change occurred (although I can make no claims to a complete causal account here). The change, regardless of its causes, has important implications for the crafting of institutional support for responsible science in the twenty-first century that have yet to be fully grasped, much less implemented.

I will begin with a description of how scientists have thought about their freedoms and their responsibilities in the period after World War II. During this time, scientific freedom and responsibility, particularly freedom from external control over their research agendas and moral responsibility[1] for the social impacts of science, were thought to be in opposition to each other. In other words, the more freedom one had, the less responsibility, and vice versa. This conception of scientific freedom was made plausible by the developing science policy context of the time, in which the linear model for science policy and a clear distinction between pure and applied science predominated. Starting in the twenty-first century, this understanding began to change, and now we have a different understanding – that scientific freedom must come with social responsibility. However, this new understanding is not yet fully implemented or institutionalized. Embracing it (as I think we should) has important implications for science policy and science education.

In what follows, I will focus primarily on freedom to set research agendas (without external planning, the key concern for scientists discussing freedom in science) and responsibility for the social impact of choices regarding which projects to pursue in science. There are many other responsibilities in science (e.g., responsibilities to colleagues, students, evidence, and subjects), but this is not the place to detail them, although their pervasive importance means they will appear in the historical narrative (see Shamoo and Resnik 2003 and Douglas 2014a for more comprehensive accounts).

In addition, I will focus on individual responsibility and individual freedom in science – rather than collective responsibility or communal responsibility. I will argue that we cannot do without individual responsibility and freedom, even if collective responsibility and communal responsibility are also important. We might decide in some circumstances to collectivize responsibility, thus partially offloading it from the individual to a designated collective (as we do with human subject oversight committees). We might decide that the scientific community as a whole is the more appropriate location for some responsibilities.[2] In general, however, some responsibility will always rest with the individual scientists. Communal responsibilities need to be acted upon by individuals; collective responsibilities need responsible individual actors to perform their roles properly. I do not have the space to delineate what should be ideally individual vs. collective vs. communal responsibilities here. I do argue that we need to craft better institutional support for the ability of scientists to meet their social responsibilities, particularly those that continue to rest with the individual.

4.2 Freedom vs. Responsibility (1945–2000)

Just as World War II was beginning, an international debate about the purpose of science and the nature of scientific freedom was brewing. In 1939, J.D. Bernal published *The Social Function of Science*, in which he argued against an ideal of pure science for its own sake and called on scientists to work together for the betterment of humanity (Bernal 1939). The publication of this book alarmed some scientists in the UK, who took it as an assault on the freedom of scientists to pursue knowledge for its own sake (Nye 2011). In 1940, John R. Baker and Michael Polanyi formed the Society for Freedom in Science (SFS) in order to counter this line of thinking from Bernal and other Marxist-leaning scientists (McGucken 1978). For leaders in the SFS, any external effort to direct scientific work toward particular societal problems was unacceptable (Douglas 2014b; Polanyi 1962). The Society grew slowly during the war, and recruited scientists in part by arguing that freedom of science was necessary to counter the authoritarian dictators Britain was fighting.

In 1944, the SFS reached out to Percy Bridgman in the US, believing he would be sympathetic to their ideas. He had already defended in print the ideal of pure science, an essential part of the SFS platform (Bridgman 1943). Bridgman became a key proponent of the freedom in science movement in the US (Bridgman 1944). As policy debates over how to fund science in the US post-war context heated up, Bridgman argued for the ability of scientists to be the sole decision makers on which projects to fund, in contrast to the Kilgore Bill, which would have required geographic equity considerations and issues of social utility to be part of decisions regarding the distribution of US federal research funds (Kleinman 1995). The debates over science funding instantiated a new science policy framework, one which utilized a clear distinction between pure and applied science, and placed science in a "linear model". In the linear model, government poured money into basic or pure research. Scientists would use these funds and themselves decide which research projects to pursue and how to pursue them, with no central planning. Basic or pure science was to be unfettered. The linear model, articulated in Vannevar Bush's 1945 monograph *Science – The Endless Frontier*, argued that investments in basic science would eventually have societal payoffs, as basic became applied research, and with application would come social benefit (thus justifying public expenditure [Bush 1945]). The essential freedom of scientists to choose their own research projects was granted for the pursuit of pure science, which was to be funded without specific social restrictions or considerations.

It was in the midst of these debates about scientific freedom and research planning[3] that Percy Bridgman (1947) wrote his classic statement on freedom and responsibility, "Scientists and Social Responsibility".[4] In it, Bridgman set forth the understanding of freedom as being opposed to responsibility (particularly responsibility for the effects of scientific work). The idea is that the more freedom one has, the less responsibility, and vice versa. It was a zero-sum game in Bridgman's view, and one that he thought should be decided in favor of complete freedom for scientists, both for decisions on what to pursue and from thinking about the social impacts of research. Indeed, he viewed with alarm growing calls for scientists' societal responsibility. Bridgman noted that in the aftermath of the development and use of nuclear weapons, the general public had "displayed a noteworthy concern with the social effects of [scientists'] discoveries" (Bridgman 1947, 148). He argued that calls for scientists to embrace moral responsibility for the effects of science would "make impossible that specialization and division of labor that is one of the foundation stones of our modern industrial civilization" and would impose a special and unique burden on scientists, because no one else is responsible for all the effects of their work (Bridgman 1947, 149). More importantly, this burden would entail a "loss of freedom" for scientists. This would undermine the value of science, according to Bridgman, because "the scientist cannot make his [sic] contribution unless he is free, and the value of his contribution is worth the price society pays for it" (1947, 149). Bridgman rejected the idea that scientists have special competencies that make them particularly qualified to grapple with the effects of science, noting the (by then) failure of scientists to establish a National Science Foundation (legislation would be disputed and delayed for another few years). But most importantly, according to Bridgman, responsibility was an anathema to freedom: "The challenge to the understanding of nature is a challenge to the utmost capacity in us. In accepting the challenge, man [sic] can dare to accept no handicaps" (1947, 153). For Bridgman, any attempt to impose social responsibility considerations on scientists was such a handicap and was tantamount to a removal of freedom to select research agendas and to pursue scientific ideas wherever they might lead.

The linear model and Bridgman's model of freedom vs. responsibility together provided for freedom of research (at least for pure scientists who are only concerned with uncovering new truths) along with a rejection of responsibility for the uses of that pure research. It was only in the application of pure science (or basic science) that one could speak of societal responsibility. Those who applied research to particular ends, who used basic research for particular purposes, bore responsibility for the impacts of science.[5] Freedom in science meant both freedom from external control of research agendas and freedom from bearing responsibility for the choices made regarding research agendas.

This view of scientific freedom and scientific responsibility was predominant for the remainder of the twentieth century. It insulated scientists from the responsibility for the *impacts* of science on society. At the same time, concerns about responsibility for the *methods* of science would become potent in the post-World War II era, with the revelations of Nazi scientific atrocities and the enshrinement of human rights protections within the context of scientific research.[6] Neither Bridgman nor the SFS addressed these kinds of moral responsibilities for scientists. Within the context of the linear model, to impose moral restrictions on scientists for methodologies seemed to be compatible with the freedom from responsibility for the downstream effects of science, and indeed compatible with freedom of scientists to decide which knowledge is worth pursuing (how to pursue it was another issue). Thus, even as controversies about scientists' responsibilities to human subjects[7] grew (and the legislation to regulate human subject research was introduced around the world – e.g., in 1974, the US passed the National Research Act), the structure of the Bridgman view on freedom and responsibility remained unchallenged.[8]

The potency of this conception can be seen in the 1975 report from the AAAS Committee on Scientific Freedom and Responsibility (Edsall 1975). The Committee had been formed in 1970, at the height of concern over Nixon's abuse of science advice (including preventing scientific advisors from speaking to Congress about particular issues and lying to Congress about the content of advising reports) and concern over the freedom of scientists in the authoritarian Soviet Union (Edsall 1975; von Hippel and Primack 1972). In addition to these concerns about government abuse of scientists, the broadly beneficial nature of science had come under sharp scrutiny in the 30 years since Bridgman's argument. In the aftermath of World War II, the development of atomic weapons could be seen as a wartime aberration from the usual course of science providing societal benefit. By 1970, such a sequestering of harm to wartime could no longer be maintained. It was not just the development of poison gases in World War I, atomic weapons in World War II, or weaponized herbicides in Vietnam that raised concerns. There were also peacetime industrial processes that contaminated the environment with DDT and other pesticides, the threat of the sonic boom from super-sonic transport, the problem of nuclear safety from peaceful nuclear power, and the growing concerns about the possible uses of genetic engineering. Science was increasingly seen as a double-edged sword, even in times of peace, and there was some consternation among scientists about what to do about this.[9] What were scientists' responsibilities and freedoms?

Despite the rejection of the blanket assumption that science was usually societally beneficial, the work at this time did not reject the framework of the linear model for discussing scientists' responsibilities. Bridgman's opposition model for freedom and responsibility continued to enjoy predominance. At first, it seemed that the AAAS Committee was open to rejecting Bridgman's oppositional approach, writing in the opening pages:

> The Committee concluded, early in its deliberations, that the issues of scientific freedom and responsibility are basically inseparable. Scientific freedom, like academic freedom, is an acquired right, generally approved by society as necessary for the advancement of knowledge from which society may benefit. The responsibilities are primary; scientists can claim no special rights, other than those possessed by every citizen, except those necessary to fulfill the responsibilities that arise from the possession of special knowledge and of the insight arising from that knowledge.
>
> (Edsall 1975, 5)

Responsibility, being primary, seems to come with scientific freedom – at least that is one way to read this passage. Yet, for the remainder of the report, the Bridgman model predominated. The Committee divided science into basic and applied for discussions of freedom and responsibility. For basic science, the central responsibilities were to maintain scientific integrity, give proper allocation of credit to other scientists, and treat human subjects appropriately (i.e., the standard Responsible Conduct of Research list). In general, basic scientists were to be given the freedom to follow their research wherever it leads (Edsall 1975, 7), and to not have a general responsibility for thinking about societal impact.

Even when discussing clearly controversial research, the Committee was reluctant to suggest that scientists were responsible for the impact of their work. For example, the Committee did note the growing concern about recombinant DNA techniques, and the efforts by scientists to grapple with these concerns, including potential moratoriums. They wrote: "Clearly this declaration [of scientists to temporarily restrict research] represents a landmark in the assumption of scientific responsibility by scientists themselves for the possible dangerous consequences of their work" (Edsall 1975, 13–14). While the Committee thought it wise to "refrain temporarily from further experiments" when those experiments might be a direct threat to human health, they were much more reluctant to consider restrictions on the freedom to do research when that research threatened "human integrity, dignity, and individuality," concerns raised with cloning (Edsall 1975, 14). They also rejected calls for an end to research into the genetic basis for individual differences in IQ (particularly when correlated with "race") (Edsall 1975, 15). In general, the Committee argued that the pursuit of knowledge should proceed even when profound ethical worries about the knowledge being produced were present. Restrictions on basic research were only to be found in the clear ethical demands for the protection of human subjects and prevention of immediate threats to human health.

For applied science, responsibility for societal impacts was a far more pressing concern. Here the Committee was much clearer in calling for general social responsibility, arguing "[m]any schemes that are technically brilliant must be rejected because their wider impact, on the whole, would be more damaging than beneficial" (Edsall 1975, 26). Further, the Committee argued that those working in applied science had much less freedom than those working in basic science, having their efforts directed by the institutions in which they worked (Edsall 1975, 27). Applied scientists thus had less freedom over their research agendas (being directed by the laboratory in which they worked, either governmental or commercial), and also more responsibility to consider the social impact of their work. In short, the Committee still maintained Bridgman's idea that the more freedom one had, the less responsibility, and the more responsibility, the less freedom, particularly regarding the direction of the research agenda.

While the AAAS Committee report represents the thinking of the mainstream of the scientific community, some scientists had decided in previous decades that their work should represent a stronger sense of responsibility. Thus, organizations arose like Pugwash (begun in 1957) and the Union of Concerned Scientists (founded in 1969), which enabled scientists to directly encourage and facilitate the pursuit of socially beneficial work by scientists. Notions such as von Hippel and Primack's "public interest science" (which was initially focused on science advising for the public interest) also helped to mark pathways to embracing greater social responsibility (von Hippel and Primack 1972; Krimsky 2003). But it was not argued that all scientists needed to embrace this sense of social responsibility, that scientists should all be trying to benefit the broader society. This was something some scientists could choose to do, but was not obligatory.

By the 1980s, some academics challenged Bridgman's view of the responsibilities of scientists (e.g., Lakoff 1980), while others (e.g., Lübbe 1986) reinforced it. Most work for the remainder of the century claimed that moral responsibilities to consider the societal impact of one's work came only with special circumstances. Thus, if one was doing applied research, if one was dealing with human or animal subjects, if one was grappling with a process that posed an immediate human health threat, or if one was involved with giving advice to the government, then such moral responsibilities might obtain. But for most scientists pursuing basic research, there were no general societal responsibilities. This can be seen, for example, in the 1992 US National Academy of Science (NAS) report *Responsible Science*, which was responding to the charge to examine the role of institutions in "promoting responsible research practices" (NAS 1992, 22). The report focused almost entirely on issues of research misconduct (i.e., data fabrication, falsification, and plagiarism). Responsible research at that time did not generally include societal responsibilities.

This is reflected as well in the NAS booklet *On Being a Scientist*, an educational booklet for the budding scientist. First published in 1989, the booklet focused on then standard philosophy of science topics (e.g., questions of scientific method, epistemic values in science, the assessment of hypotheses) and internal research ethics questions (e.g., fraud and error, apportioning credit for discovery, peer review). A brief section at the end addressed broader societal responsibilities of scientists. After quickly noting the importance of protecting human subjects, animal subjects, and the environment, and the prevalence of societal responsibility in applied science, it said:

> Scientists conducting basic research also need to be aware that their work ultimately may have a great impact on society. ... The occurrence and consequences of discoveries in basic research are virtually impossible to foresee. Nevertheless, the scientific community must recognize the potential for such discoveries and be prepared to address the questions that they raise.
>
> (NAS 1989, 9072)

Examples like the temporary moratorium on recombinant DNA were touted as exceptional demonstrations of societal responsibility (NAS 1989, 9072). In addition, the booklet noted that some scientists take on roles that have more social responsibility (such as governmental science advising), but that other scientists prefer not to be involved in such efforts. The booklet upheld the freedom of the individual scientist to choose whether to take on explicit social responsibility, even if "dealing with the public is a fundamental responsibility for the scientific community" (NAS 1989, 9073). Someone needed to interact with the public to maintain public trust, but plenty of scientists could simply opt out. Thus continued the model of societal responsibility only arising under special circumstances, and otherwise being optional.

When the NAS revised the booklet for a second edition (1995), this treatment of societal responsibilities in science remained mostly unchanged. The booklet repeats much of what was written in 1989, but added that "If scientists do find that their discoveries have implications for some important aspect of public affairs, they have a responsibility to call attention to the public issues involved" (NAS 1995, 20–21).

Here we see the beginning of a new, general responsibility for all scientists, starting off as a responsibility to communicate the results of their research when relevant to public interest. (This was a reflection of the growing call for scientists to communicate their research relevant to climate change, for example.) Scientists still had general freedom to choose their basic research agenda, and freedom *from* responsibility for the social impacts of their choices.

This division between responsibility and freedom was not restricted to the US. It can also be seen in the structures of the International Council of Scientific Unions (ICSU). Work on the freedom of movement of scientists (particularly to attend scientific meetings internationally) began in 1965, and by 1996, this committee shifted attention from freedom of movement to freedom in science more generally, becoming the Standing Committee on Freedom in the Conduct of Science. Also in 1996, a new committee, the Standing Committee on Responsibility and Ethics in Science (SCRES), was formed to address issues of responsible conduct of science. The two committees operated independently of each other (Schindler 2009).

In the 60 years after Bridgman had proposed that freedom and responsibility were antagonistic, an accumulating list of exceptions to this rule (not for applied science, not for human subjects, not for science advisors, etc.) had helped to protect the idea that, at the heart of science and generally, scientists should not be responsible for the societal impact of their work. Responsibilities for societal impact were limited to special circumstances, and could sometimes be offloaded to institutional oversight (such as IRBs for human subject research or IACUCs for animal research). But it was not considered a responsibility of scientists to actively attempt to shape their research agendas, at least when doing "basic research", to avoid societal harms or to provide societal benefits. This was still seen as part of the protection of scientific freedom for research agenda-setting from external control. Thus, by the end of the twentieth century, the expectation of general societal responsibility for scientists was still limited.

4.3 Freedom with Responsibility

In the first decade of the twenty-first century, however, the Bridgman model would finally be overturned. One cause for the rejection of the idea that freedom and responsibility were in a zero-sum tension – that freedom meant in part freedom from responsibility – was the increasing untenability of all the exceptions required to maintain it. The growing list of all the exceptions to the claim that freedom to choose research agendas also meant freedom from social responsibility for the impacts of science made the space for such a kind of freedom increasingly elusive. That the model was the proper general understanding for freedom and responsibility in science looked untenable with all the exceptions. But it was the rising concerns over dual-use research that made clear that Bridgman's freedom vs. responsibility model was not the right model. Such concerns eliminated any remaining space where Bridgman's model could still hold sway.

Dual-use research (i.e., research that could be put to unintended harmful uses) became a prominent concern after the September 11, 2001 attacks and the anthrax mailing attacks later that fall. That there were malicious non-state actors and that scientific research could be used for harmful purposes by such actors became a recognized backdrop for research, particularly in the biomedical sciences. Initially, some scientists grappling with policy for dual-use research tried to clearly define domains where it would be a problem, so that, by implication, other areas of research need not think about dual-use possibilities. But this attempt proved unworkable, as dual-use concerns cropped up in all kinds of places (in and out of the biomedical sciences).

The importance of individual scientists' responsibility for grappling with dual-use concerns was apparent to many from the start. In a 2002 joint statement from the presidents of the UK Royal Society and the US NAS, they argued that:

> Individual scientists also have a key role to play. Every researcher, whether in academia, in government research facilities, or in industry, needs to be aware of the potential unintended consequences of their own and their colleagues' research. In 1975, scientists agreed to the "Asilomar moratorium," which gave guidance to researchers in the emerging field of recombinant DNA research. Today, researchers in the biological sciences again need to take responsibility for helping to prevent the potential misuses of their work, while being careful to preserve the vitality of their disciplines as required to contribute to human welfare.
>
> (Alberts and May 2002)

Similarly, in the 2004 US National Research Council (NRC) report on *Biotechnology Research in an Age of Terrorism*, the first recommendation was to improve education about and discussion concerning dual-use potential across biotech research. Yet the report also attempted to map out where additional oversight might be called for, in an attempt to show when scientists had to (and did not have to) think about dual-use issues. The report produced seven categories of "experiments of concern":

1   Demonstrate how to render a vaccine ineffective
2   Confer resistance to therapeutically useful antibiotics or antiviral agents
3   Enhance the virulence of a pathogen or render a non-pathogen virulent
4   Increase the transmissibility of a pathogen

5   Alter the host range of a pathogen
6   Enable the evasion of diagnosis and/or detection by established methods
7   Enable the weaponization of a biological agent or toxin

(NRC 2004)

Three years later, in a comprehensive discussion of dual-use research, Miller and Selgelid felt compelled to add four more categories:

8   Genetic sequencing of pathogens
9   Synthesis of pathogenic micro-organisms
10   Any experiment with variola virus (smallpox)
11   Attempts to recover/revive past pathogens

(Miller and Selgelid 2007)

Today, even this expansive set would need to be added to. While the H5N1 gain-of-function experiments would clearly fall within this net and be properly labeled dual-use research, the 2017 reconstruction of horsepox may not (and in fact initially did not because it is not a human pathogen) – and yet it too raised potent dual-use concerns because of the similarities between horsepox and smallpox (Enserink 2011; Kupferschmidt 2018). Further, these categories are only for the biomedical fields. Dual-use concerns arise in many other fields, such as nanotechnology (e.g., self-replicating nanobots, nanoweaponary), number theory (e.g., cryptography), and AI (e.g., lethal autonomous weapons, the generation of fake news footage, and state surveillance).

In addition, not all cases of potentially problematic research arise from dual-use concerns, as reference to Asilomar in the quote from Alberts and May above makes clear. Some research is potentially dangerous from accidents, some from possible intended (and harmful) uses, and some from unintended impacts that are not clearly accidents. Geoff Brumfiel discussed just four cases of areas of research concern in *Nature* in 2012 (radioisotope separation, brain scanning, geoengineering, and fetal genetic screening), noting that these examples "are hardly a definitive list, but they do give a sense of how frequently such conundrums arise – and show that scientists must constantly ask themselves whether the benefits [of a research project] outweigh the risks" (Brumfiel 2012, 432). As cases of morally problematic research agendas have multiplied, it is clear that no field or subject is in principle immune from legitimate ethical and societal concerns over the selection of research agendas.

In retrospect, we should not be surprised by our inability to define ahead of time the areas of research where these moral issues would arise. The history of nuclear weapons development shows clearly how quickly an area of "pure research" could become practical and morally fraught. In the summer of 1938, nuclear physics was an area of pure research with no foreseeable practical application. With the discovery of fission that fall (and the enormous release of energy from a fission event), and the subsequent confirmation of multiple neutrons coming from each fission event (allowing for the possibility of chain reactions), nuclear physics research was suddenly at the center of moral quandaries. The response of physicists in 1939 to this development as the world hurtled toward World War II was to debate internally what should (and what should not) be openly published and how to get the important developments in front of the right governments (Rhodes 1986). This example shows that the specter of dangerous or risky research can arise suddenly, and scientists must be ready to act responsibly when confronting such developments – where what counts as responsible action depends a great deal on the broader geopolitical and social context.[10]

By 2010, these considerations caused a shift from the Bridgman oppositional model between freedom and responsibility to seeing responsibility as a necessary partner to scientific freedom, as yoked to freedom. No longer was freedom to set research agendas thought to be only secured when one also was free from social responsibility for the impacts of research. Instead, freedom to pursue research was now thought to require bearing responsibility for the impacts of research. Such a shift appears in the official documents and structures of the scientific community. An early sign of this shift is found in ICSU's policy shifts. In 2006, ICSU disbanded the separate committees on freedom and responsibility and reformed them with new members as one committee, the Committee on Freedom and Responsibility in the Conduct of Science (CFRS).

In the resulting ICSU document, *Freedom, Responsibility and Universality of Science* (2008), emphasis was placed on

> not only freedoms of scientists, including freedom of movement, freedom of association, freedom of expression and communication, and access to data, information and research materials, but also scientists' responsibilities for the conduct of science and their responsibilities to society.
>
> (Schindler 2009)

In the 2014 revision of the pamphlet, the freedom of scientists is placed in necessary contact with responsibility for science, including "respect and care for physical, social and human environments". The pamphlet recognizes that

> Not all applications of science are beneficial for everyone or the environment. The use of scientific knowledge can pose serious social or physical threats. Therefore, scientists as individuals and the international scientific community have a shared responsibility, together with other members of society, to do their utmost to assure that scientific discoveries are used solely to promote the common good.
>
> (ICSU 2014, 5)

The idea that the pursuit of science is automatically generally good for society has been left behind. Yet the vestiges of the old Bridgman model are still present in the pamphlet, when it talks about "tensions between freedoms and responsibilities" for scientists (ICSU 2014, 20).

The third edition of *On Being a Scientist*, published in 2009, took a much stronger stand on social responsibility:[11]

> The standards of science extend beyond responsibilities that are internal to the scientific community. Researchers also have a responsibility to reflect on how their work and the knowledge they are generating might be used in the broader society.
>
> (NAS 2009, 48)

This is not just under certain circumstances nor just about communicating results. The responsibility is for *all* scientists and concerns how any result might be used, i.e., its ultimate societal impact.[12]

More recent statements present even stronger positions that reject the Bridgman model. For example, the 2017 AAAS Statement on Scientific Freedom and Responsibility states:

> Scientific freedom and scientific responsibility are essential to the advancement of human knowledge for the benefit of all. Scientific freedom is the freedom to engage in scientific inquiry, pursue and apply knowledge, and communicate openly. This freedom is inextricably linked to and must be exercised in accordance with scientific responsibility. Scientific responsibility is the duty to conduct and apply science with integrity, in the interest of humanity, in a spirit of stewardship for the environment, and with respect for human rights.
>
> (AAAS 2017)

Here we have a full official expression of societal responsibility yoked to scientific freedom. They are seen as intertwined – that one cannot have the freedom to select research agendas and to pursue research without bearing the responsibility for the impacts of those choices, including ensuring that research is done "in the interest of humanity". There are no caveats here – this does not apply only to scientists doing applied research, or only with respect to the need to avoid unethical methodologies. All scientists who expect freedom must also embrace social responsibility.[13]

Or consider the 2018 World Economic Forum (Young Scientists) Code of Ethics, which states:

> It is the responsibility of every scientist to both consider the possible consequences of their research practices, outcomes and publications, and to undertake such research according to ethical principles. Scientists may not always have control over the findings or the end use of their research, but this does not absolve them from the responsibility to make a sincere effort to bring about positive change for society and their professional community.
>
> (WEF 2018)

The responsibility to consider the impacts of scientific research on society fall on each and every scientist, regardless of any attempt to pursue basic or applied research, or in which field they are working.[14]

In short, the freedom to set one's research agenda as a scientist, once thought to require freedom from responsibility for the societal impacts of one's choices (Bridgman's model), now is seen as requiring a corresponding responsibility for such impacts. If scientists want to be free to choose their research agendas, they need to be responsible for those choices as well. The freer scientists are, the more responsibility they will have.

4.4 Making it Work in Practice

What does this changing conception of the relationship between freedom and responsibility mean for scientific practice and scientific governance? Is this workable in practice?

Let me first address Bridgman's concern from 1947, that bearing responsibility for the impact of their work would completely hamstring scientists. Recall that Bridgman worried that if any and all societal impacts were the responsibility of individual researchers, researchers would spend all their time trying to forestall negative impacts of their past work rather than doing more research. I think this concern can be set aside when we recognize that the scope of responsibility Bridgman presumed is too broad. Bridgman thought that bearing social responsibility for the setting of the research agenda meant that *all* societal impacts of novel research would be the past responsibility of the individual scientist. But this is too expansive, and would indeed place a special burden on scientists, distinct from what all moral actors bear and requiring insight that would be superhuman. We can follow Douglas (2003) and argue instead that general moral responsibility does not encompass all consequences, but rather all reasonably foreseeable consequences or impacts. Note that this is broader than what might be intended by the scientist, but narrower than *all* impacts. Further, a scientist might reasonably judge that some risk of harmful effects is worth the possible benefits (including consideration of the distribution of effects).

What counts as reasonably foreseeable is something that is likely to be contested and contestable. Nevertheless, we can say some things about how it should work. It should be judged from the perspective of what is known at the time a decision is made, and not from hindsight. It is not fair to judge past responsibility with the full knowledge of the present. It should be judged drawing from cross-disciplinary expertise, as Lefevere and Schliesser (2014) have noted. And I think it should be judged on the basis of clearly plausible mechanisms known at the time rather than wild speculation. For example, backwards time travel and perpetual motion machines remain in the realm of the not plausibly foreseeable (even if we can tell fictional stories about them), while widespread deployment of autonomous vehicles and sea-level rise from climate change are plausibly foreseeable. When there are difficult cases, one of the things we will debate is whether there was plausible foreseeability and thus social responsibility.

If societal responsibility for individuals is curtailed in this way, Bridgman's argument that responsibility would hamstring scientists and be an unfair burden is undercut. The requirement to avoid being negligent is one that all people share and encourages careful forethought by scientists, without requiring perfect foresight (an impossible standard). So the responsibility that comes with freedom is neither unusual for scientists nor out of reach. It requires more from scientists than a simple belief that no moral responsibility for the impacts of science belongs to them, however. What we need to address is how to properly support scientists in the bearing and meeting of this responsibility.

Attending to how to fully implement this understanding of responsibility with freedom is made further pressing when one realizes that the view of responsibility as a necessary partner for freedom is not yet generally institutionalized. We are still living with the hangover from the struggles with Bridgman's understanding of freedom vs. responsibility, which brought with it an effort to carve off those situations where responsibility encroached. Our default institutional approach to social responsibility in science is still to attempt to define precisely when (what was once special) social responsibility arises, and then create regulatory frameworks to provide oversight for those situations (e.g., Institutional Review Boards for human subject research, Institutional Animal Care and Use Committees for animal research, special oversight triggers such as Select Agent Lists) combined with checklists to maintain regulatory compliance. The attempts to create compliance-based ethical systems, where scientists can meet a checklist of requirements and thus be adequately responsible, is a vestige of the old way of thinking about freedom and responsibility. Such checklists attempt to offload responsibility to some other actor, so that scientists need only think about whether they have met predetermined criteria before proceeding. But there is no predetermined criteria we can articulate to govern science for responsible research. A key characteristic of scientific research is that unexpected results can arise at any time. It is the individual scientist (or the particular collaborative team) that first encounters the unexpected. They must act responsibly in the face of the unexpected, and stay attuned to the moral issues that can arise in the midst of research. Scientists' social responsibility can thus never be fully offloaded to others.

Further, individual scientists do not yet fully embrace the social responsibility that should come with freedom of research. Many are still resistant to it being their job to always think of such things, and still seek to find a space where freedom exists without responsibility.[15] But compliance-based, checklist ethics has never served science, nor society, well. As Sabina Leonelli (2016) argued (in the context of big data), there is no checklist that will adequately manage the development of new modes of research. Scientists, if they are to be free to pursue new research avenues, must also be responsible for their choices.

This means we should move away from a strict governance model for dual-use research concerns and for responsible research in general. Science education should teach responsible practice of science as something that threads throughout the doing of science, not something that one only has to think about at certain points. There is no checklist that will ensure scientists are properly responsible.

We can still have oversight for areas where there are clear minimum standards we want to enforce and well-known moral complications (e.g., human subject research, biosafety laboratory standards). Doing so should be thought of as at most a partial off-loading of responsibility, through collectivization. When we collectivize responsibility, we designate a group of actors – a group that often should still include the scientific researcher pursuing the work – to bear the responsibility for a difficult decision. The group as a whole (rather than the individuals in the group) is then responsible for the decision, and also holds the freedom and authority to make the key decision.[16] With collectivized responsibility, freedom and responsibility go together still; it is just the group that is now both free to make the choices and responsible for the foreseeable impacts of the choices. Because collectivizing responsibility means a loss of freedom for individual scientists, we do this sparingly, only when clear need presents itself. Where responsibility has not been collectivized, freedom and responsibility remain with the individual researcher.

Thus, in general we should shift the attention of scientists from compliance to full responsibility in their decision-making. This does not mean, however, that scientists need to embrace their responsibility alone. Rather than trying to define ever more fine-grained contexts when compulsory oversight is warranted, we should instead set up places where scientists can go to ask for help with the thorny ethical issues they will inevitably encounter. We developed the idea of an ethical consultant for medical practice in the 1970s and 1980s and it is now standard in hospitals to have such an option for physicians. It is time to import this idea to research practice. Scientists should have clear and known places to call for help when they need to decide whether to pursue or to publish a particular piece of research, to help think through the implications of various courses of action and to make a responsible decision. Advisory bodies such as the NSABB and dual-use advisory committees for journals are already in place for the most prominent cases.[17] But lower-level advising mechanisms, such as ethics consultation for scientists or, more intensively, STIR protocols,[18] can help scientists navigate their freedoms and the attendant responsibilities.

Having a standard institutional location at universities, as well as in private and government laboratories, to go to for an ethics consultation would be a huge help. Science education should also normalize the idea that thinking about ethical implications is a regular part of doing scientific work. This is crucial if the joint nature of freedom and responsibility is to thrive in practice.

4.5 Conclusions

In the aftermath of World War II, the scientific community embraced an unworkable conception of scientific freedom and social responsibility as oppositional. In the twenty-first century, scientific societies, driven by problems with the earlier model and growing concerns about the pervasiveness of dangerous and/or dual-use research, have moved to a more plausible model, where freedom comes with responsibility, not for all impacts of research, but for the plausibly foreseeable ones.

There is fuzziness around this set of downstream effects in practice. What one person will think is plausibly foreseeable will not be thought so by others. It thus behooves scientists to remain in discussion with colleagues (both within and beyond their disciplines) about the plausible impact of their work. Further, institutional support (rather than oversight) for these decisions should be a standard aspect of every research institution. Discussions of the possible impact of work should not just be throwaway lines in a grant proposal, but carefully thought about and discussed aspects of shaping research agendas. Because scientists are responsible not just for what they intend but also for plausible impacts, forethought must be part of the work of being a responsible scientist. All scientists must attend to the question over whether they *should*, not just whether they *can*.

Curtailing the freedom of scientists through central planning is not a good way to get at the societal responsibilities of scientists. The Society for Freedom in Science was right about this, but Bridgman was wrong that scientists shed their societal responsibility for their choices about which research projects to pursue. Providing institutional support for scientists' decision-making regarding their research agendas and scientists' reflection on the societal implications of their choices is a better approach.

Ultimately, freedom and responsibility together must rest with the individual scientist. Even if the scientist chooses to collaborate with the public, to consult with colleagues, to place issues before an advisory board (like the NSABB), the scientist still[19] has the freedom to pursue research as they see fit and bears the responsibility for those choices. Given both the intractability of full oversight for scientists' decisions and the importance of responsibility for those decisions, it is doubtful we could have science any other way.

Acknowledgements

My thanks to Péter Hartl for inviting me to participate in the July 2019 conference on *Science, Freedom, Democracy* at the Hungarian Academy of Sciences, when I received helpful feedback from the group in Budapest (despite giving the talk via videoconference). An earlier version of this work was given in 2018 at the *Scientific Governance at the Ground Level Workshop*, at the Centre for the Study of Existential Risk, University of Cambridge. In August

2019, I gave a revised version at the *Congress of Logic, Methodology, and Philosophy of Science and Technology* (CLMPST) in Prague and am grateful for the audience's questions. Ted Richards, Joyce Havstad, my colleagues in the Socially Engaged Philosophy of Science (SEPOS) group at MSU, and several reviewers provided crucial assistance refining the expression of these ideas. Remaining missteps are my responsibility.

Notes

1 For this chapter, I will use the term "social responsibility" to denote moral responsibilities that we have with respect to the societies in which we function. Social responsibility is thus the requirement to think about the broader societal implications of one's choices, beyond the impacts on oneself and on one's family, friends, and colleagues.

2 For example, it is usually a communal responsibility to communicate the results of science effectively, and thus not the responsibility of each individual scientist to do so. But individual scientists who do communicate with the media, for example, must communicate effectively and responsibly if the community is to meet its communal responsibility.

3 See Reisch (2019) for a rich examination of the US debates around scientific freedom, planning, and science policy in the 1930s, 1940s, and 1950s, from Unity of Science to Kuhn.

4 It was first delivered at the American Association for the Advancement of Science (AAAS) December 1946 meeting and reprinted in *Bulletin of the Atomic Scientists* in 1948.

5 This was a tacit understanding of how to think about research policy for the SFS, but the linear model and Bridgman's essay made it explicit.

6 Such concerns did not originate with the Nuremberg trials. As Horner and Minifie (2011) show, such concerns date back at least to the beginnings of the nineteenth century. See also Lederer (1995).

7 There is a similar history of growing concern and increasing legislation regarding animal subject research. See Fuchs (2000) for an overview.

8 Indeed, the SFS disbanded in 1963, partly because the battle for freedom seemed mostly won, but also partly because of a dispute over John Baker's eugenicist views (Reinisch 2000).

9 See also the collection of talks at the time, *Scientists in Search of their Conscience* (Michaelis and Harvey 1973).

10 Contemporary examples of how new discovery can cause rapid changes in scientists' sense of societal responsibility – because what becomes plausibly foreseeable changes – include the discovery of CRISPR, which allows for far more precise genetic engineering, and the development of neural net AI, which can allow for rapid and complex AI learning. Voluntary (i.e., generated by the scientific community itself) restrictions of research agendas have been proposed for both (regarding human genome editing and lethal autonomous weapons development, respectively). The debate continues, and we should be grateful such debates are not happening in the midst of the beginning of world war.

11 The booklet still treats these responsibilities very briefly after the usual topics such as data handling, research misconduct, credit allocation, and a new section on intellectual property issues, reflecting growing concerns about the commercialization of science.

12 In addition to these developments in the documents of scientific societies, commentators on responsible science made stronger calls for general responsibilities

for scientists (e.g., Douglas 2003) or for stronger role responsibilities for scientists (e.g., Mitcham 2003). It is ironic that the subtitle for my 2003 paper, "Tensions between Autonomy and Responsibility", reflects the older Bridgman model even as I argued against such a model. It also does not seem that these academic arguments caused the change in views.

13 Douglas (2003) argues for this view, although in terms of autonomy and responsibility, rather than freedom and responsibility.

14 In the EU, developing ideas around responsible innovation and the Horizon 2020 reflect this shift as well. See, e.g., Owen, Macnaghten, and Stilgoe (2012).

15 One can see such fears expressed in surveys such as Carrier and Gartzlaff (2020) or in demarcation rationality in Glerup and Horst (2014).

16 This is already a standard part of grant decision-making in federal agencies, where a committee is tasked with deciding which grants should be funded and not. The responsibility for such decisions is thus collectivized, i.e., not resting with a particular individual but shared by a designated group.

17 Journals such as *Science, Nature,* and *PLOS One* already have such committees. Salsbury (2011) was an early discussion of the need for these kinds of oversight mechanisms.

18 The STIR protocol embeds a social scientist or humanist in a lab for a few months to ask questions about the research agenda and methods, and to raise ethical issues and the possibilities for alternative research avenues. It has a track record of improving the productivity of labs as well as improving the beneficial nature of the work done. It also can help to retrain scientists to think about their responsibilities that come with their freedoms more regularly. See, e.g., Schuurbiers and Fisher (2009); Fisher and Schuurbiers (2013).

19 In most cases – fully illegal research agendas exist (such as offensive biological or chemical weapons research) but such restrictions are rare.

References

Alberts, B., and R.M. May. 2002. "Scientist Support for Biological Weapons Controls." *Science* 298 (5596): 1135–1136.

American Association for the Advancement of Science. 2017. "AAAS Statement on Scientific Freedom and Responsibility". Available online at www.aaas.org/pa ge/aaas-statement-scientific-freedom-responsibility.

Bernal, J.D. 1939. *The Social Function of Science*. London: G. Routledge & Sons.

Bridgman, P.W. 1943. "Science, and its Changing Social Environment." *Science* 97: 147–150.

Bridgman, P.W. 1944. "The British Society for Freedom in Science." *Science* 100: 54–57.

Bridgman, P.W. 1947. "Scientists and Social Responsibility." *The Scientific Monthly* 65 (2): 148–154.

Brumfiel, G. 2012. "Good Science/Bad Science." *Nature* 484 (7395): 432–434.

Bush, V. 1945. *Science: The Endless Frontier*. Washington, DC: US Government Printing Office.

Carrier, M., and M. Gartzlaff. 2020. "Responsible Research and Innovation: Hopes and Fears in the Scientific Community in Europe." *Journal of Responsible Innovation* 7 (2): 149–169.

Douglas, H. 2003. "The Moral Responsibilities of Scientists: Tensions between Autonomy and Responsibility." *American Philosophical Quarterly* 40 (1): 59–68.

Douglas, H. 2014a. "The Moral Terrain of Science." *Erkenntnis* 79 (5): 961–979.

Douglas, H. 2014b. "Pure Science and the Problem of Progress." *Studies in History and Philosophy of Science Part A* 46: 55–63.

Edsall, J.T. 1975. *Scientific Freedom and Responsibility*. Washington, DC: AAAS.

Enserink, M. 2011. "Controversial Studies give a Deadly Flu Virus Wings." *Science* 334: 1192–1193.

Fisher, E., and D. Schuurbiers. 2013. "Socio-technical Integration Research: Collaborative Inquiry at the Midstream of Research and Development." In *Early Engagement and New Technologies: Opening up the Laboratory*, edited by N. Doornet al., pp. 97–110. Dordrecht: Springer.

Fuchs, B.A. 2000. "Use of Animals in Biomedical Experimentation." In *Scientific Integrity*, edited by F. Macrina, pp. 101–130. Washington, DC: ASM Press.

Glerup, C., and M. Horst. 2014. "Mapping 'Social Responsibility' in Science." *Journal of Responsible Innovation* 1 (1): 31–50.

Horner, J., and F. Minifie. 2011. "Research Ethics I: Responsible Conduct of Research (RCR) – Historical and Contemporary Issues Pertaining to Human and Animal Experimentation." *Journal of Speech, Language, and Hearing Research* 54: 303–329.

ICSU. 2014. *International Council for Science: Freedom, Responsibility and Universality of Science*. Available online at https://council.science/wp-content/uploads/2017/04/CFRS-brochure-2014.pdf.

Kleinman, D. 1995. *Politics on the Endless Frontier*. Durham, NC: Duke University Press.

Krimsky, S. 2003. *Science in the Private Interest*. Oxford: Rowman and Littlefield.

Kupferschmidt, K. 2018. "Critics See Only Risks, No Benefits in Horsepox Paper." *Science* 359 (6374): 375–376.

Lakoff, S.A. 1980. "Moral Responsibility and the 'Galilean Imperative.'" *Ethics* 91 (1): 100–116.

Lederer, S.E. 1995. *Subjected to Science: Human Experimentation in America Before the Second World War*. Baltimore, MD: Johns Hopkins University Press.

Lefevere, M., and E. Schliesser. 2014. "Private Epistemic Virtue, Public Vices: Moral Responsibility in the Policy Sciences." In *Experts and Consensus in Social Science*, edited by C. Martini and M. Boumans, pp. 275–295. Cham: Springer.

Leonelli, S. 2016. "Locating Ethics in Data Science: Responsibility and Accountability in Global and Distributed Knowledge Production Systems." *Philosophical Transactions of the Royal Society A: Mathematical, Physical and Engineering Sciences* 374 (2083): 20160122.

Lübbe, H. 1986. "Scientific Practice and Responsibility." In *Facts and Values*, edited by M.C. Doeser and J.N. Kray, pp. 81–95. Dordrecht: Springer.

McGucken, W. 1978. "On Freedom and Planning in Science: The Society for Freedom in Science, 1940–46." *Minerva* 16 (1): 42–72.

Michaelis, A.R., and H. Harvey. 1973. *Scientists in Search of Their Conscience*. New York: Springer.

Miller, S., and M.J. Selgelid. 2007. "Ethical and Philosophical Consideration of the Dual-use Dilemma in the Biological Sciences." *Science and Engineering Ethics* 13 (4): 523–580.

Mitcham, C. 2003. "Co-responsibility for Research Integrity." *Science and Engineering Ethics* 9 (2): 273–290.

National Academy of Sciences (NAS), Committee on the Conduct of Science. 1989. "On Being a Scientist." *Proceedings of the National Academy of Sciences of the United States of America*: 9053–9074.

National Academy of Sciences (NAS), Institute of Medicine, National Research Council, & Panel on Scientific Responsibility and the Conduct of Research. 1992. *Responsible Science: Ensuring the Integrity of the Research Process.* National Academies Press.

National Academy of Sciences (NAS). 1995. *On Being a Scientist: A Guide to Responsible Conduct in Research.* National Academies Press (US).

National Academy of Sciences (NAS). 2009. *On Being a Scientist: A Guide to Responsible Conduct in Research.* National Academies Press (US).

National Research Council (NRC). 2004. *Biotechnology Research in an Age of Terrorism.* National Academies Press.

Nye, M.J. 2011. *Michael Polanyi and his Generation.* Chicago: University of Chicago Press.

Owen, R., P. Macnaghten, and J. Stilgoe. 2012. "Responsible Research and Innovation: From Science in Society to Science for Society, with Society." *Science and Public Policy* 39 (6): 751–760.

Polanyi, M. 1962. "The Republic of Science." *Minerva* 1 (1): 54–73.

Reinisch, J. 2000. "The Society for Freedom in Science, 1940–1963." MSc Diss. London: University of London.

Reisch, G.A. 2019. "What a Difference a Decade Makes: The Planning Debates and the Fate of the Unity of Science Movement." In *Neurath Reconsidered: New Sources and Perspectives*, edited by J. Cat and A.T. Tuboly, pp. 385–411. Cham: Springer.

Rhodes, R. 1986. *The Making of the Atomic Bomb.* New York: Simon and Schuster.

Salsbury, D. 2011. "Editors Must be Aware of Dual-use Research." *Science Editor* 34 (3): 97.

Schindler, P. 2009. "A Short History of the Committee on Freedom and Responsibility in the Conduct of Science (CFRS) and its Predecessor Committees of the International Council for Science (ICSU)." Available online at https://council.science/wp-content/uploads/2017/04/CFRS_history.pdf.

Schuurbiers, D., and E. Fisher. 2009. "Lab-scale Intervention." *EMBO Reports* 10 (5): 424–427.

Shamoo, A.E., and D. Resnik. 2003. *Responsible Conduct of Research.* Oxford: Oxford University Press.

Von Hippel, F., and J. Primack. 1972. "Public Interest Science." *Science* 177 (4055): 1166–1171.

World Economic Forum. 2018. *Young Scientists Code of Ethics.* Geneva: World Economic Forum. Available online at www3.weforum.org/docs/WEF_Code_of_Ethics.pdf.

5   Bacon's Promise

Janet A. Kourany

DEPARTMENT OF PHILOSOPHY, UNIVERSITY OF NOTRE DAME

5.1 The Promise

At the dawn of modern science, a promise was made. If society would but support the new enterprise, society would be richly rewarded not only with unprecedented insights into the workings of the universe but also with all the benefits such insights would provide. Indeed, Francis Bacon, one of the chief architects of the new science as well as one of its more exuberant press agents, promised that the knowledge science would offer would "establish and extend the power and dominion of the human race itself over the universe" for the benefit of all humankind (Bacon 1620/1960, 117–119). What did Bacon mean? The problem, as he saw it, was that the human race had been thrust into "immeasurable helplessness and poverty" by the Fall from Eden, and needed to be rescued. And science would be the rescuer. In other words, science would provide a solution to the plight of humankind (Bacon 1603/1964).[1]

To explain how this would go, Bacon offered a blueprint for the new science, a blueprint that was later adopted by the Royal Society as well as other early scientific societies and that is still in effect today. In it he included a variety of illustrations of the benefits to humankind he expected of the new science. Science, he suggested, would make possible the curing of diseases and the preservation and prolongation of life, science would produce the means to control plant and animal generation, science would lead to the development of new building and clothing and other kinds of materials, science would provide new modes of transportation ("through the air" and "under water") and even new modes of defense (Bacon 1627/2008). In all these ways and others too, science would make humans once again the masters of nature as they had been in the Garden of Eden, and hence once again "peaceful, happy, prosperous and secure" (Bacon 1603/1964).

Essentially the same promise was made again and again over the next four centuries by other distinguished representatives of the scientific establishment, though typically without the theological trappings Bacon included. For example, one of these reiterations – probably the most famous – was the one put forward by Vannevar Bush, the engineer and inventor who headed the US Office of Scientific Research and Development (OSRD) during the Second World War. At the end of that war, Bush sent a report to President Franklin D. Roosevelt that became the basis of US science policy for much of the twentieth century. In it Bush promised that, if science is supported by society but also left free of societal control, its advances will bring

> more jobs, higher wages, shorter hours, more abundant crops, more leisure for recreation, for study, for learning how to live without the deadening drudgery which has been the burden of the common man for ages past. Advances in science will also bring higher standards of living, will lead to the prevention or cure of diseases, will promote conservation of our limited national resources, and will assure means of defense against aggression.
>
> (Bush, 1945, 10)

Moreover, Bush added, such advances in science will be crucial for attaining these benefits: "Without scientific progress no amount of achievement in other directions can insure our health, prosperity, and security as a nation in the modern world" (1945, 11).

So, here was Bacon's promise again. The seventeenth-century theological infusions were gone, to be sure, but so much else remained. Indeed, where Bush now promised "health, prosperity, and security" for people as a result of science, Bacon had promised that they would be "peaceful, happy, prosperous and secure" as well as healthy; where Bush now promised that science would banish the "deadening drudgery" of their pre-science existence, Bacon had promised that science would end the "immeasurable helplessness and poverty" of that existence; and so on.

Bush's promise did depart from Bacon's in one respect, however. It had to do with what counted as *legitimate* science and how social benefits would arise from it. For Bacon, scientific research was all about – *should be* all about – attending to the needs of society:

> Lastly, I would address one general admonition to all – that they consider what are the true ends of knowledge, and that they seek it not either for pleasure of the mind, or for contention, or for superiority to others, or for profit, or fame, or power, or any of these inferior things, but for the benefit and use of life, and that they perfect and govern it in charity. For it was from lust of power that the angels fell, from lust of knowledge that man fell; but of charity there can be no excess, neither did angel or man ever come in danger by it.
>
> (Bacon 1620/1960, 15–16)

If such utility-driven research were supported, Bacon promised, science's social benefits would result. For Bush, on the other hand, the most important kind of scientific research, the kind on which all other scientific research depends, was all about freely pursuing "the truth wherever it may lead". "Scientific progress on a broad front results from the free play of free intellects, working on subjects of their own choice, in the manner dictated by their curiosity for exploration of the unknown" (Bush, 1945, 12). And only if society supported *that* kind of research would science's social benefits result.

By the end of the twentieth century, however, "the free play of free intellects" was no longer considered "the best precondition for maximizing the utility of science" (Rohe 2017, 745; for more details, see, for example, Gibbons 1999; Guston 2000; Krishna 2014; Sarewitz 2016). Science had just gotten too big and too costly, with no end in sight to its continued and ever-increasing demands for support. As a result,

> The sheer size of the system and its need for sustainable allocation of funds is finally unbalancing Bush's claim for the "free play of free intellects". ... To continue feeding the science system, a broad societal consensus is needed, in which legitimization is increasingly, often tightly, linked to performance measures and other demonstrable evidence of contributions to social welfare, economic growth, and national security.
>
> (Rohe 2017, 746)

No matter. Whether the free play of free intellects was what yielded the social benefits of science (as Bush had claimed) or whether they resulted most reliably only from research explicitly aimed at them (as Bacon had suggested), Bacon's promise – that such benefits *would* result if science were supported – was still very much taken for granted. But after four centuries of support for science, has that happened? Has Bacon's promise been kept?

Certainly, science has provided, or helped to provide, such things as food in ever greater variety and abundance, produced more quickly and efficiently; the near eradication of such dreaded diseases as scarlet fever, smallpox, and polio, and impressive progress on other diseases such as HIV/AIDS; better insulated, more comfortable homes, with more conveniences, produced more quickly and efficiently; more sophisticated communications systems; and quicker, more convenient modes of transportation. And in the future even more extraordinary benefits are expected, such as tiny, inexpensive computers that are thousands of times more powerful than current machines, flying automobiles and other kinds of vehicles that help us multitask, and human lives that are nearly disease free and last for 150 years or more. All this is precisely the kind of outcome Bacon had promised.

But science has also provided, or helped to provide, such things as a food supply tainted with every manner of pesticides, herbicides, antibiotics, growth hormones, and other harmful chemicals; polluted air and water and the looming menace of global warming; ever-rising mountains of garbage and toxic wastes; ever more prevalent heart disease and strokes, cancer, diabetes, gallbladder disease, and other dreaded diseases related to unhealthy (fat-filled, sugar-filled, salt-filled, calorie-filled) diets and polluted environments; ever more depleted supplies of the world's resources and widespread extinction of plant and animal life; and, of course, enormous stockpiles of nuclear and other weapons.

What's more, the benefits that science *has* provided have improved the lot of only some of humankind, not all of humankind. Indeed, scientific investigation has largely ignored the needs of many in the developed world and nearly all in the developing world. Medical research, for example, has devoted more than 90% of its resources into problems that affect only 10% of the world's population. Left out of research are

> diseases that predominantly affect developing countries (the "neglected diseases"), ... the specific needs of developing countries in relation to diseases with a global incidence, and ... the development of affordable medicines for all. But the problem of neglect extends beyond the developing world, as becomes clear from the global lack of R&D for new antibiotics, appropriate children's medicines (and other products), and orphan diseases.
>
> (Viergever 2013)

Moreover, scientific investigation may even be helping to intensify the needs of those who have been ignored. "The experience of the past 30 or more years shows that the phenomenon of science-and-technology-based economic growth seems to be accompanied by increasing inequality in distribution of economic benefits" (Sarewitz et al. 2004, 69).

> Paradoxically, across major economies, while new technologies have boomed, productivity growth has declined, slowing the main engine of economic growth. Income and wealth inequalities have risen. The proverbial economic pie has been growing more slowly and more unequally, feeding today's social discontent and political divisiveness.
>
> (Qureshi 2019)

And there is every indication that this inequality accompanying science-and-technology-based development will continue to increase:

> Inequality as a result of technological innovation isn't a forgone conclusion, but it's clear that society as a whole needs to get much better at improving the skills development of all citizens if the dividend is to be spread more widely. Sadly, there is little evidence that governments even understand this dilemma, much less are actively looking to address it.
>
> (Gaskell 2019)

In short, after four centuries of societal support for science, Bacon's promise regarding the social benefits of science is still unfulfilled. Of course, there are many possible explanations for why this is so.

One explanation, for example, concerns the object of scientific investigation. According to this explanation, the world and the challenges it poses are far more complex than Bacon and the other founders of the new science ever imagined – far too complex to expect any uniformly positive results to be achievable by science.

A second explanation, closely related to the first, concerns the method used by science to investigate the world. According to this explanation, science's inductive method, or the human mind that makes use of this method, is much weaker than Bacon and the others supposed – too weak to ensure the kind of results from science that Bacon had promised.

Yet a third explanation concerns the institutions that over time have come to support or collaborate with or in other ways influence science. Perhaps science *could* have achieved all that Bacon and the others had expected had it been left to itself, but according to this explanation, science has been corrupted by outside interests. Many today, for example, complain that science has been commercialized and politicized and militarized and compromised by cultural biases such as sexism and racism, and that this is why its results have been as mixed as we find them.

But a fourth explanation, by contrast, denies both that science was originally good and that it is now corrupted. Rather, science and the information science makes available are simply resources that can be used for good or ill (for example, by industry, government, or the military) and, not surprisingly, they *have* been used for good *and* ill. So again, no uniformly positive results should have been expected from science.

Doubtless there are still other explanations for why Bacon's promise regarding the social benefits of science remains unfulfilled, and doubtless there is some truth in all of them. What is striking about these explanations, however, is that they all seem to let *scientists* off the hook. It is the complexity of the world and its challenges or the limitations of the inductive method or the human mind, it is the outside interests that have corrupted science or the uses that have been made of science, that bear the blame for the mixed results of science over the past centuries. Scientists themselves aren't to blame, aren't to be held accountable. But granting whatever truth lies within these explanations, can't scientists also do better than they have, shouldn't they do better, and shouldn't we strive to help them do better? These are the questions I should like to press.

5.2 A Peek at the Way Science Has Been

It helps to get concrete. Consider, then, the scene in the United States, the place I know best. There people enjoy a vast array of comforts and conveniences – in housing, food, clothing, transportation, medical care, you name it – thanks, in large part, to science. But people there also suffer from a multitude of health problems. Even after nearly 50 years of War-on-Cancer research costing many billions of dollars, one out of every two men and one out of every three women will develop cancer during their lifetimes. Although there has been a continuous decline in the cancer death rate for about three decades now, still, an estimated 606,520 Americans will die from it in 2020 – 1,600 people every day (Siegel, Miller, and Jemal 2020). At the same time, heart disease continues to be the leading cause of death among both women and men, killing more Americans than all forms of cancer combined (Virani et al. 2020). And obesity has now reached epidemic proportions, with more than two out of every three adults and one out of every three children either overweight or obese, conditions that have been linked to heart disease as well as diabetes and stroke and a variety of other problems (National Institute of Diabetes and Digestive and Kidney Diseases 2017). What's more, there are respiratory diseases such as emphysema and bronchitis, Alzheimer's disease, kidney disease, neurodevelopmental disorders such as autism and ADHD – and this is just the tip of the iceberg. But these health problems of Americans, like the comforts and conveniences we enjoy, are also thanks, in large part, to science. How can this be?

Consider, to begin with, chemistry. The twentieth century in the United States has been called the Synthetic Century. By the end of that century around 80,000 synthetic chemicals were in use in the US – in pesticides and fertilizers, in clothing and furniture, in cars and personal care products and pharmaceuticals, and of course in the nearly universal accompaniment of modern life, plastic. In roughly the last quarter of the century alone, the country's consumption of synthetic chemicals had increased 8,200%. At the same time, many more synthetic chemicals were being released into the environment as wastes from the manufacture of these 80,000 chemicals. But only a tiny fraction of all these chemicals (7%) had been fully tested for toxicity – have been fully tested, even today – and only a tinier fraction (1%) have been tested for their effects on human health. Even so, many of these chemicals are now being linked to such health problems as autism and learning disabilities, Alzheimer's disease, Parkinson's disease, diabetes, asthma, decreased fertility, and birth defects, as well as cancer and heart disease (Cranor 2011; Jenkins 2011; Landrigan and Landrigan 2018).

Four choices on the part of twentieth-century chemists, chemical engineers, and chemical industry executives were pivotal to this sorry state of affairs (see, for what follows, Thornton 2000; Stevens 2002; Woodhouse 2003; 2006). The first of these choices was to focus on certain raw materials or "feedstocks" from which to synthesize chemicals – in particular, petroleum-based feedstocks. The second choice was to rely on wet chemical processes of synthesis rather than dry ones – that is, chemical processes that require solvents, many of which (such as benzene and toluene) are extremely toxic. The third choice was to consider as acceptable processes of synthesis that involve many separate reactions (as many as 30), each of which produces byproducts, that is, chemicals not usable in later parts of the process. These chemical byproducts were sometimes used in other processes of synthesis, but sometimes they were not usable at all – that is, were simply wastes, and frequently hazardous wastes. Such multi-staged processes have produced millions of tons of hazardous wastes. Finally, the fourth choice was to aim for a rapid transition from initial laboratory synthesis through pilot plants to full-scale production. This "rapid scale-up", as it is called, left little time to consider the safety of releasing the new chemicals into ecosystems and human environments.

Were there alternatives to these four choices? Apparently there were. Although petroleum-based feedstocks were cheap, plentiful, and reliable, other feedstocks were available as well, such as woody plant materials and fatty/oily animal and plant materials. And the chemistries needed to deal with such feedstocks – carbohydrate chemistry and lipid chemistry – were already well established. Work on dry synthesis was also available and such syntheses would have saved money in a variety of ways (e.g., no solvents or post-synthesis solvent removal would have been needed). Insisting on simpler processes that generated less waste to dispose of with more useable product would also have saved money. Of course, there were also economic reasons to make the choices that were made – for example, to aim for rapid scale-up and for products (such as DDT, PCBs, dieldrin, polyvinyl chloride, and other chlorinated hydrocarbons) that would use up the large quantities of chlorine that were byproducts of the manufacture of caustic soda (that is, lye, sodium hydroxide), a core product of the chemical industry.

Still, warning signs of problems began to accumulate early on – obvious ones such as massive die-offs of bees after fruit trees were sprayed with the chemicals produced as well as subtle ones such as the thinning of eggshells. But even after Rachel Carson's trenchant criticism (1962) and the rise of the environmental movement in the middle of the twentieth century, there was little change in the choices of the chemical establishment. Did chemists, in particular, have a free hand to make other choices? Certainly chemistry was and still is more closely tied to industry and industry's needs – more *captive to industry*, as some have put it – than perhaps any other science. And certainly, as Rensselaer Polytechnic Institute political scientist Edward Woodhouse has emphasized, "chemists did not act alone": "industry actually manufactured the chemicals, and millions purchased them". Still, chemists "created the potential that set the agenda" (Woodhouse 2006, 169). And more important, chemists continue to set that same agenda today. "Some of the world's brightest and most highly trained experts continue to poison their fellow humans and the ecosystem without fundamental reconsideration of whether there is a better way to do things" (Woodhouse 2006, 155).

But chemistry is not the only science that helped to produce the widespread health problems now ailing the US (see, for what follows, Moss 2013). The twentieth century also saw the birth of convenience foods and fast foods and snack foods and junk foods. Even baby boomers who once considered it unthinkable not to sit down to at least two home-cooked meals a day with perhaps one planned snack at bedtime became hooked on such foods, and their children and grandchildren have often known little else. Meanwhile, in universities and research institutes, such as the Monell Chemical Senses Center in Philadelphia, as well as in the laboratories of companies such as Kraft Foods and Frito-Lay, legions of food scientists – physiologists and chemists and neuroscientists and biologists and psychologists and geneticists – carefully engineered these processed foods to make them irresistible.

Indeed, everything about these foods was engineered – the shape and size and internal structure of the salt granules used, the distribution and texture of the fat, the way it melts in the mouth (called "mouth-feel"), the concentration of sugar to create the most powerful response of enjoyment (called the "bliss point"), package size and design and location in convenience stores and grocery food aisles, the language used on the packages ("toasted" and "baked," for example, not "fried"), even the creation of a social environment in which it is okay to eat everywhere and all the time – everything about these foods was engineered to compel consumption and overconsumption. The result was an American diet loaded with salt, sugar, fat, and calories. By the end of the twentieth century, for example, Americans were eating so much salt that they were getting 10 times, even 20 times, the amount of sodium their bodies needed. And high blood pressure and the diseases associated with high blood pressure as well as obesity and diabetes were sharply on the rise.

True, many well-placed food scientists were becoming concerned about the growing obesity problem in the United States and their role in producing it, and many criticized the industry because of it. Scientists at the Monell Chemical Senses Center, for example, despite the substantial financial support the Center accepted from the largest food companies, took exception to the way food manufacturers used sugar to increase the allure of their products. And industry scientists even tried, at times, to redirect their companies' research priorities, and sometimes the whole industry's research priorities, toward a healthier alternative. Still, these scientists honored the constraints set by the industry – to make no changes, even changes that increased the healthfulness of their products or at least decreased their harmfulness, that diminished the products' appeal or the profits they delivered. What's more, these scientists, when the needs of competition demanded it, were even willing to further compromise the public's health and wellbeing. As a result, quite a number have now come to regret the work they did or failed to do. Confided one recently – Robert Lin, former chief scientist at Frito-Lay – "I feel so sorry for the public" (quoted in Moss 2013, 306).

Consider one last area of US science, medical research. With all the health problems visited on the US population thanks to sciences such as chemistry and food science, it is crucial that American medical research conscientiously do its part to ameliorate the situation. Unfortunately, however, American medical research hasn't seemed to do that. For one thing, researchers have tended to focus on genetic causes and the biological mechanisms involved in conditions such as cancer and heart disease rather than the environmental causes and lifestyle choices that chemistry and food science make salient. For another thing, researchers have tended to neglect most of the subgroups of the US population when developing treatments for such conditions even when these subgroups are at least equally in need of the treatments.

Consider only the second shortcoming with regard to just heart disease, and consider, in particular, the situation of women. Heart disease is the leading cause of death among women. In fact, in the US one woman dies from heart disease every minute (Virani et al. 2020). Yet, until the 1990s heart disease was studied primarily in white, middle-aged, middle-class men. The large, well-publicized, well-funded studies of the past are illustrative: the Physicians' Health Study, whose results were published in 1989, examined the effect of low-dose aspirin therapy on the risk of heart attack in 22,071 male physicians; the Multiple Risk Factor Intervention Trial (MR FIT), whose results were published in 1990, examined the impact of losing weight, giving up smoking, and lowering cholesterol levels on the risk of heart attack in 12,866 men; the Health Professionals Follow-Up Study, whose results were also published in 1990, examined the relation of coffee consumption and heart disease in 45,589 men. And these studies were no exceptions: in a 1992 *Journal of the American Medical Association* analysis of all clinical trials of medications used to treat acute heart attack published in English-language journals between 1960 and 1991, for example, it was found that fewer than 20 percent of the subjects were women (Rosser 1994).

The consequences of this neglect of women were profound. Since women were not researched along with the men, it was not discovered for years that women differed from men in symptoms, patterns of disease development, and reactions to treatment. As a result, heart disease in women was often not detected and it was often not even suspected. What's more, it was not properly managed when it was detected (see, for details, Weisman and Cassard 1994; Schiebinger 1999).

All this was to be rectified in 1993 when Congress passed the National Institutes of Health Revitalization Act, which mandated the inclusion of women and minority men in NIH-funded US medical research. But as it turns out, the Revitalization Act had serious limitations. For one thing, it did not apply to early-phase medical studies. In consequence, most basic research with animal models continued to focus on male animals and exclude females, most studies at the tissue and cellular levels either failed to report the donor sexes of the materials studied or reported that they were male, and early phase clinical trials (that is, Phase I and II trials) didn't always include women. Even in the case of the medical studies to which the Revitalization Act did apply (that is, Phase III clinical trials, the final stage after the earlier, smaller trials indicated safety and promise), women remained under-enrolled relative to their representation in the patient population, and the published results frequently did not include a breakdown by gender. What's more, the Revitalization Act applied only to NIH-funded research, not the sizeable amount of biomedical research funded by either private industry or foundations, and most of that research also failed to live up to the standards set by the Revitalization Act.

Ultimately, the problem was that many in the biomedical research community simply did not support the goals of the Revitalization Act. There were, of course, the old standby reasons – that including women of child-bearing age in drug studies could jeopardize the health and safety of any fetuses the women might be carrying, that men and women were so alike, anyway, that results obtained from studying men could always be validly applied to women, and finally (and quite inconsistently with the latter) that women, with their menstrual cycles, oral contraceptives, hormone therapies, pregnancies, and so on, would introduce too many complications to make "clean" results possible if they were included in the studies. But there was also the bottom line: that including women in clinical trials, or including female animals, tissues, or cells in earlier stage studies, would require much larger studies, more expense, and more work if done properly – that is, so as to ensure that an adequate analysis of results by sex and gender would be possible. And, in any case, there was the inertial resistance to changing old, entrenched procedures as well as the recognition that even the relatively weak requirements of the Revitalization Act were not being rigorously monitored and enforced even for NIH-funded research (Johnson et al. 2014; Mazure and Jones 2015; Liu and Dipietro Mager 2016; Dusenbery 2018).

The upshot?

> Steps have been taken in the United States to remedy the under-representation of women and the inadequate attention to sex and gender differences in research and regulatory approvals. However, progress has been painfully slow – stalling for long periods or sometimes reversing direction – and, consequently, not nearly enough progress has been made.
>
> (Mazure and Jones 2015, 2)

Meanwhile, the effects of the pre-1993 exclusions still linger. For example, women consume roughly 80 percent of the pharmaceuticals used in the United States, but they are still frequently prescribed drugs and dosages devised for men's conditions and average weights and metabolisms. As a result, adverse reactions to drugs occur twice as often in women as in men. "The net effect of gender bias in medical research and education", concludes historian of science Londa Schiebinger, "is that women suffer unnecessarily and die. ... Not only are drugs developed for men potentially dangerous for women; drugs potentially beneficial to women may [have been] eliminated in early testing because the test group [did] not include women" (Schiebinger 1999, 115).

5.3 A Peek at the Way Science Can Be

Chemistry, food science, and medical research, then, have failed to live up to Bacon's promise. But the reason for that failure does not seem especially to lie with any of the explanations offered at the outset, the explanations, you will recall, that functioned to exclude scientists from responsibility. At least, the reason for the failure does not seem to involve either the complexity of the world or the weakness of the inductive method or the human mind, the first two of the possible explanations offered. After all, even far more detailed accounts of the events I have described have yet to yield a picture of scientists at a conceptual loss, struggling against insurmountable odds. Nor does the reason for Bacon's broken promise seem to lie with the neutrality of science, the neutrality that would allow science to be used for both good and ill – which was the fourth explanation put forward. Although some science may be neutral in this way, the science I described – research into ways of making hazardous chemicals to be used in hazardous products; research into ways of producing compulsive eating of unhealthful foods; and research into ways of dealing most effectively only with certain groups' health problems – the science I described does not seem to merit the "neutral" label.

If any of the four explanations offered at the outset fits our cases, it would be the third explanation, concerned with the corrupting of science by outside interests – in the cases at hand, commercial and cultural interests. But this explanation also has its problems. True, probably a higher percentage of scientists in chemistry and food science work for industry than is the norm in science, and hence probably a higher percentage than the norm are vulnerable to possible industry pressures. And yet, the evidence we have – for example, the extensive interviews of food scientists provided by Pulitzer Prize-winning *New York Times* investigative reporter Michael Moss in his recent book *Salt, Sugar, Fat: How the Food Giants Hooked Us* (and we have nothing comparable regarding the history of chemistry in the twentieth century) – the evidence we have suggests that industry scientists were not only very much at one with industry goals but were centrally involved with shaping those goals. And as for the university and research institute scientists supported by industry, the evidence suggests that they pursued the questions that most interested them, knowing full well all the while (whether they liked it or not) that their results would be used for less-than-laudable industry goals (see, for example, Moss's illustrations of the "tension between the industry's excitement about the research at [the Monell Chemical Senses Center in Philadelphia] and the center's own unease about the industry's practices" (Moss 2013, 8). So, at least these cases provide no support for the view that scientists or their science (its content or its culture) are hapless victims of external industry pressures. Finally, although medical research has been located in an androcentric and sexist culture throughout the twentieth and now the twenty-first centuries, the androcentrism and sexism that has characterized the research has been introduced by the scientists themselves or their discipline's traditional practices rather than by some external representative of the culture. So, again, the failure to live up to Bacon's promise seems to lie very much with the scientists and their science rather than something external to that science.

Of course, American chemists, food scientists, and heart disease researchers were not alone in their failure to live up to Bacon's promise during the twentieth century. It is notorious that American physicists failed as well, and so did many others, such as the social scientists and biologists whose work added to and reinforced traditional biases not only against women, but also against people of color, members of the LGBTQ community, the poor, and the disabled. Nor were all these scientists the only relevant actors involved. The American public, for example, eagerly embraced all the wonders and conveniences afforded us by twentieth-century science, and we reveled in the new opportunities and influence that science gave us on the world stage. Still, the science that brought us the wonders and the opportunities also fostered much of our failure to understand the problems that accompanied these wonders and opportunities (think, for example, of the ignorance and deception as well as enjoyments engineered by food science). And now, in the twenty-first century we may be headed for yet another round of failures to keep Bacon's promise – this time on the part of artificial intelligence and robotics researchers as well as practitioners of the other emerging technosciences.

The story of Bacon's promise is not all negative, however, even with regard to the sciences on which I have focused. Some scientists *have* lived up to it, or at least they have tried. In the case of twentieth-century chemistry, for example, a small minority of chemists did conduct research outside the petrochemical mainstream – for example in carbohydrate chemistry and lipid chemistry (see, for example, the work described in Stevens 2002) – and in 1973 Stanford University chemist Barry Trost proposed the principle of "atom economy" as a response to both the limited availability of raw materials and the need to safeguard the environment. According to this principle, chemical processes should be judged not by whether they produce a usable product at a satisfactory cost – the mainstream standard – but by whether they achieve the highest possible percentage of input atoms in the usable output, ideally leaving no waste. This principle, together with 11 others concerned with such topics as the use of renewable feedstocks and safer solvents or no solvents at all, was later elaborated in Paul Anastas and John Warner's groundbreaking 1998 book, *Green Chemistry: Theory and Practice*, the first comprehensive treatment of environmentally responsible chemistry (Anastas and Warner 1998; Woodhouse 2006).

Warner and Anastas, in fact, have been credited with founding the new field of green chemistry, and Warner, in 2001, founded the world's first green chemistry PhD program at the University of Massachusetts, the same university that houses the Lowell Center for Sustainable Production, the research and advocacy center that works to apply the principles of green chemistry to industry. There are now comparable centers in a variety of places, including Yale University and the University of California, Berkeley in the US and the University of York in the UK.

Some food scientists have also tried to live up to Bacon's promise. An especially noteworthy such scientist was twentieth-century British physiologist and nutritionist John Yudkin, who founded what became an internationally acclaimed nutrition department at the University of London's Queen Elizabeth College, setting his sights on improving the health of the public through nutrition. Alarmed by the growing incidence of heart disease during the first half of the twentieth century, Yudkin, along with other nutritionists both in Europe and the United States, looked to diet as a possible cause of the problem. But while Yudkin, early on (see, for example, Yudkin 1957; 1964; 1972), placed the blame on sugar in particular, noting the ever-increasing amounts of sugar in twentieth-century diets, most other nutritionists blamed the high amounts of fats in those diets, especially saturated fats (see, for example, the work of American epidemiologist Ancel Keys, an important spokesman for the fat side of the debate and a vocal critic of Yudkin – see, especially, Keys 1971; 1975; 1980).

Of course, both the advocates of sugar as the cause of the heart disease and the advocates of fat as the cause had their evidence, epidemiological and experimental, and they both had their critiques of the other side's evidence. But – more important, as it turns out – the advocates of the fat hypothesis also had the backing of the sugar industry, which, starting in the 1960s, generously funded research that downplayed the dangers of sugar and highlighted the dangers of fat, and even *de*funded research that threatened to do the opposite (Kearns, Schmidt, and Glantz 2016; Nestle 2016; Kearns, Apollonio, and Glantz 2017). Yudkin, in contrast, had the *hostility* of the sugar industry to deal with. "Jobs and research grants that might predictably have come Yudkin's way did not always materialize", conferences he organized that the sugar industry suspected would promulgate anti-sugar findings were abruptly cancelled, and Yudkin's most important work linking sugar to heart disease was viciously attacked (Watts 2013, 17). In the process, Yudkin's scientific reputation was

> all but sunk. He found himself uninvited from international conferences on nutrition. Research journals refused his papers. He was talked about by fellow scientists as an eccentric, a lone obsessive. Eventually, he became a scare story [of what would happen if a scientist dared to critique sugar].
>
> (Leslie 2016)

Nonetheless, Yudkin carried on with his anti-sugar message until his death in 1995, even though by then his adversaries had triumphed and their fat hypothesis had become a central principle of national and international dietary guidelines.

The story does not end there, however. For other scientists are now taking up where Yudkin left off. University of California, San Francisco, pediatric endocrinologist Robert Lustig, for example, has not only continued research and outreach to the public on the dire health effects of sugar but has also emphasized the impossible challenges to the healthcare system and the economy these effects pose (see, for example, Lustig 2009; 2012). And he has been instrumental in the reissue of Yudkin's 1972 book *Pure, White, and Deadly: The Problem of Sugar*, which he says was not only completely correct but also prophetic of what future research would disclose. Meanwhile others, such as New York University nutrition researcher Marion Nestle, are exposing the ways the sugar industry as well as the other food industries continue to manipulate food science in their favor (see, for example, her 2018 book *Unsavory Truth: How Food Companies Skew the Science of What We Eat*). Unfortunately, such manipulation, Nestle points out, is not likely to end while government and foundation grants are scarce and scientists are forced to rely on industry funding to support their research. So, the struggle within food science continues.

Feminist medical research furnishes further examples of scientists who tried to fulfill Bacon's promise. Its founders, in particular, not only worked to bring about equal attention to women's and men's needs in medical research, but they also began the difficult but necessary reconceptualization of what such equality of attention requires. Indeed, medical researchers such as Society for Women's Health Research founder Florence Haseltine presided over a shift in women's health research from merely reproductive research (involving attention to childbirth, contraception, abortion, premenstrual syndrome, breast and uterine cancers, and the like) to more general health research (involving attention to women's distinctive physiology), and this has been critical to improving health care for women. Other medical researchers, in turn – such as Adele Clarke, Elizabeth Fee, Vanessa Gamble, and Nancy Krieger – moved the understanding of health research toward a broader social model that does more than focus on disease management and biochemical processes. This broader model takes into account how health and disease are produced by people's daily lives, access to medical care, economic standing, and relations to their community. It thereby takes into account the differing health needs of women and men of different races and ethnic origins and classes and sexual orientations (see Moss 1996 and Schiebinger 1999 for more details regarding all these developments; see, also, the impressive work now being done for the Gendered Innovations in Science, Health and Medicine, Engineering, and Environment Project directed by Schiebinger (n.d.), and the nearly 100 scientists all over the world who are contributing to it).

But exactly what is it that these chemists, food scientists, and medical researchers have done that other scientists have not done but ought to have done? Elsewhere I have described it in terms of what I call the ideal of socially responsible science (see Kourany 2010). According to this ideal, sound social values as well as sound epistemic values must control every aspect of the scientific research process, from the choice of research questions to the communication and application of results. In these terms, green chemists, health-conscious food scientists, and feminist medical researchers, but not their more traditional scientific colleagues, have been striving to realize the ideal of socially responsible science.

Take green chemists. Like all chemists, green chemists study, design, manufacture, and put to use chemicals and chemical processes to meet society's basic needs for food, clothing, shelter, health, energy, clean air, water, and soil. But green chemists are also committed to not polluting the environment; not jeopardizing the safety and health of consumers, workers, and those in the communities around workplaces; and conserving energy and natural resources; as well as maintaining the long-term economic viability of industry. And they design their research with these additional social values very much in mind. Indeed, these additional social values underlie the so-called "twelve principles of green chemistry" that form the foundation of green chemists' research practice. The principles include, among others (see Anastas and Warner 1998 for the complete list):

- Design chemical syntheses to prevent waste, leaving no waste to treat or clean up.
- Use renewable feedstocks such as agricultural products or the wastes of other processes, not depleting feedstocks such as those made from petroleum.
- Avoid using solvents, separation agents, or other auxiliary chemicals, or use innocuous chemicals (such as water or supercritical carbon dioxide, that is, carbon dioxide in a state midway between a gas and a liquid) if such chemicals are necessary.
- Design chemical products to break down to innocuous substances after use so that they do not accumulate in the environment.

Traditional chemists, by contrast – so-called "brown chemists" – seem *not* to be committed in their research practice to green chemists' additional social values. Or more precisely, they seem committed to only one of them, the one about the importance of maintaining the long-term economic viability of industry. For traditional chemists make use of depleting feedstocks and hazardous solvents; they design chemical syntheses that generate lots of hazardous waste; they put chemical plant workers and consumers of their products at risk of diseases such as cancer; and so on. As a result, traditional chemists cause serious (avoidable!) harms in their pursuit of chemistry's sound social values.

Achieving socially responsible chemistry, then, requires instilling a more robust set of sound social values into the discipline of chemistry – into the codes of ethics prescribed for chemists (which right now are unbelievably meager); into the enforcement procedures supporting these codes (which right now don't exist); into the training given to future chemists (which right now typically requires competence in a foreign language but no competence in toxicology or green chemistry); into the kinds of activities and practices held up as valuable and rewarded with prizes and fellowships and research grants (which right now, with few exceptions, support brown chemistry, not green chemistry); into the criteria used for publication; and so on. Achieving socially responsible chemistry requires instilling a more robust set of sound social values into all these aspects of the discipline of chemistry. Such a change would have important benefits.

Consider just one of the benefits. We are now witnessing what some call the third wave of the environmental movement in the US. The first wave, associated with names such as John Muir and Theodore Roosevelt, focused on conservation and the fight to keep large parcels of land off-limits to commercial development. The second wave, inspired by the publication of *Silent Spring*, focused on legislation to control industrial pollution, such as the Clean Air Act and the formation of the Environmental Protection Agency. This third wave, by contrast, focuses on consumers. The idea is that corporate behavior can be changed by consumers educated about the perils of consumer products and thereby motivated to change their buying habits. Hence the flurry of recent books such as *Dodging the Toxic Bullet: How to Protect Yourself from Everyday Environmental Health Hazards* by David Boyd (2010), and *What's Gotten into Us: Staying Healthy in a Toxic World* by McKay Jenkins (2011). And hence the stepped-up educational campaigns of organizations such as the Environmental Working Group. All of this is intended to safeguard consumers' health and wellbeing but also to generate consumer pressure on the chemical industry.

But notice that, as ingenious as this third wave is, and as valuable as the second wave has been, their immediate concern is damage control – dealing with toxic chemicals already out there or soon to be out there. Changing the discipline of chemistry, its culture, by contrast, would deal with the problem at its source, preventing chemists' synthesis of toxic chemicals right from the start. What's more, it would protect individual chemists from pressures that industry might be tempted to exert to get certain kinds of results. Indeed, it would back with the whole weight of the culture – the culture of chemistry – the refusal by individual chemists to satisfy ethically unacceptable corporate demands. But it could offer other benefits as well, such as a way to deal with decreasing public trust in science and decreasing public support.

Of course, changing the discipline of chemistry – or, in fact, nearly any other scientific discipline – to make it socially responsible would be neither quick nor easy. But it may prove to be the only way finally to fulfill Bacon's promise.

5.4 The Take-Home Message

The foregoing has been a very preliminary treatment of its topic.[2] Indeed, we have merely *peeked* at Bacon's promise regarding science and the four centuries of promises that followed, and merely peeked, again, at both the way science has been and the way science can be. Nevertheless, on the basis of these mere *peeks*, a fairly radical overhaul of sciences such as chemistry, food science, and biomedical research was proposed. The prudent reader might well wonder, as a result, whether all this was not too hasty. Is there not a more cautious way to respond to these sciences? Of course, the proposal was to enable sciences such as chemistry, food science, and bio-medical research to be more socially responsible, and thereby more in keeping with Bacon's promise – hardly a dangerous plan of action. Still, perhaps we overreacted and these sciences are less in need of transformation than we supposed. There are, in fact, at least three lines of argument that suggest as much.

The first employs a cost-benefit approach. As we saw, Bacon promised, at the dawn of modern science, that science would provide abundant benefits for all of humankind if adequately supported, and over the centuries other representatives of the science establishment promised this too. Now, reflecting on all that has been accomplished by science during this time, all the wonderful comforts and conveniences and opportunities that have been afforded us by science, might we not simply conclude that science *has* ful-filled Bacon's promise? True, there have been problems – false starts, errors along the way, miscalculations, and failures of one sort or another, but looking all at once at the whole picture, haven't these been but the inevitable costs of the daring feats science has pursued? An appropriate response at this point might therefore be nothing more than admiration for what has been accomplished, unending gratitude, and the appropriate kinds of patch-ups for the occasional problems. Why won't this cost-benefit approach do?

The reason is that it fails to adequately represent the mix of costs and benefits we actually confront. As already explained, the costs currently inflicted by chemistry, food science, and biomedical research are far more severe than the occasional blemish on success that the above characterization suggests. Indeed, these costs now include serious threats to the prospects of a healthy life and even the habitability of our earthly home, and these costs have been meted out very unequally to the world's population, with some even shouldering the costs of others' benefits. Worse still, many of these costs have been knowingly produced, many even deliberately produced, and although methods are available to reduce such costs – we saw examples in the last section – they are not being adopted on a wide enough scale. So, a cost-benefit approach does not seem an especially promising strategy of evasion for the transformation of the sciences already proposed.

Turn, then, to the second line of argument. This one employs a linguistic approach. Like the cost-benefit approach, the linguistic approach offers us a way to claim that Bacon's promise *has* been fulfilled, and hence that there is really no significant fix of the sciences now needed. More specifically, the linguistic approach aims to equip us with greater clarity regarding science and its distinction from technology. According to it, *science* refers to the diverse array of activities, typically housed in universities and research institutes, that are concerned with knowledge building, while *technology* (or applied science and engineering) refers to the diverse array of activities, typically housed in industrial and government research settings, that are concerned with the *application* of this knowledge for practical purposes. Now, since scientists cannot be expected to predict all the particular applications that will be made of the knowledge they uncover once their research is completed and the knowledge is out of their control in the public domain, scientists bear no responsibility for those applications. So, shortcomings associated with those applications are not *scientists'* failings or the failings of their science, but rather the failings of the technologists who have made the applications. It follows that the failings of chemistry, food science, and biomedical research previously described – failings clearly associated with the *applications* of science – were failings of *technology*, not science, and hence not failings of Bacon's promise at all. With this clarity offered by the linguistic approach in hand, then, we have what seems a fully workable way to evade the proposed transformation of the sciences.

Unfortunately, however, this second strategy of evasion faces at least as many problems as the first. For Bacon's promise concerned an enterprise that many would now call *technoscience* (see, for the concept, Latour 1987), that is, an enterprise in which the technological and the scientific so closely interact that they constitute two components of a single set of activities rather than two separate sets of activities. And, chemistry, food science, and biomedical research, like most contemporary sciences, are such technosciences: the "scientific" activities associated with them, though they may occur in universities and research institutes, are typically funded by industry, using products and equipment provided by industry, and geared to the practical needs and ends of industry. As a result, the failings of chemistry, food science, and biomedical research that we discussed previously are relevant to Bacon's promise and, in fact, serve to document how it has not been fulfilled. Indeed, if Bacon's promise referred to what the linguistic approach defined as science – an enquiry completely divorced from the practical – then such science could hardly be expected to yield any benefits at all. For that it would need technology – in particular, helpful technology (the kind that yields benefits), not harmful technology (the kind that yields harms). And so, Bacon's promise, under this reading, would become: science will help everyone flourish if aided by a technology that will help everyone flourish – hardly an exciting promise!

This leaves the third line of argument to evade all thought of transforming the sciences in the way proposed. This third line of argument employs a no autonomy/no responsibility approach. Such an approach fully recognizes that Bacon's promise has not been fulfilled, though it quickly adds that many factors account for this. Of course, those factors might include scientists' commitments to values that are not socially responsible, but they also include, more importantly, constraints on scientists' research activities – constraints imposed by industry and/or government and/or universities and/or the public at large and/or particular interest groups. Any or all of these constraints might, and frequently do, shape scientists' research, and thus what emerges from that research, in ways that fail to promote human flourishing. They all represent curtailments of scientists' research autonomy, and hence curtailments of scientists' responsibility for the failure of Bacon's promise. Accordingly, it is simply naive to expect that merely infusing the culture of science with the "right" social values, the ones that promote human flourishing, will enable Bacon's promise to be fulfilled. Scientists just don't have the control that such a response presupposes.

The problem here is that scientists *do* have that control, at least potentially. To see this, however, we need to shift from an individual perspective to a social perspective. Think, for example, of what happens in a workplace. While individual workers may not – probably will not – be able to exert much control over their working conditions, a union representing all the workers typically *will* be able to exert that control (through, for example, the possibility of strikes and collective bargaining). Why can't the same be true of scientists? After all, few things happen without the input of scientists – certainly not the diverse and constantly changing projects of industry, or the complex undertakings of governments, or the innovations or just plain workings of universities, or the bulk of the comforts, conveniences, and enjoyments of the public. Consequently, if the scientific community *as* a community were to refuse to pursue a certain kind of research as not in keeping with Bacon's promise, or were to insist on a different kind of research as fully supporting that promise, such a stand could effectively control the research that is done and its aftermath in a socially responsible way. And the fact that the scientific community does not do this very frequently.– *has not* done this, for example, in the areas discussed in this paper – reflects poorly on their enterprise as a deliberately chosen failure to fulfill Bacon's promise. But what if the culture of that community were infused with the right social values, the ones that promote human flourishing, the ones that define Bacon's promise? As I said before, achieving this would be neither quick nor easy. But it may be the only way to fulfill Bacon's promise. At any rate, this is the take-home message that I hope you will take home.

Acknowledgements

Early versions of this paper were presented at the University of Alberta, Canada; Bielefeld University, Germany; Inter-University Centre for Postgraduate Studies, Croatia; University of Lyon, France; Zhejiang University and Nanjing University, China; and University of Notre Dame, USA. I would like to thank the audiences at these universities for their very helpful questions and suggestions and their wonderfully lively discussions. I would also like to thank Ádám Tamás Tuboly, Péter Hartl, and an anonymous reviewer for their careful reading of the paper and astute recommendations for its further development.

Notes

1 Some have thought that Bacon was disingenuous in all this, simply manipulating religious themes to support a very materialist, hedonist, purely secular agenda. See, for example, White (1968) for a prominent defense of this position, and McKnight (2005; 2006) for an effective rebuttal.
2 In contradistinction to the book it deserves that I hope to produce.

References

Anastas, P., and J. Warner. 1998. *Green Chemistry: Theory and Practice*. Oxford: Oxford University Press.

Bacon, F. 1603/1964. "The Masculine Birth of Time." In *The Philosophy of Francis Bacon: An Essay on Its Development from 1603 to 1609, with New Translations of Fundamental Texts*, edited and translated by B. Farrington, pp. 59–72. Liverpool: Liverpool University Press.

Bacon, F. 1620/1960. *The New Organon and Related Writings*, edited by F.H. Anderson. Translated by J. Spedding, R.L. Ellis, and D.D. Heath. Indianapolis and New York: Bobbs-Merrill.

Bacon, F. 1627/2008. *The New Atlantis*, Project Gutenberg EBook #2434. Available online at www.gutenberg.org/files/2434/2434-h/2434-h.htm. Retrieved January 16, 2020.

Boyd, D. 2010. *Dodging the Toxic Bullet: How to Protect Yourself from Everyday Environmental Health Hazards*. Vancouver: Greystone Books.

Bush, V. 1945. *Science, the Endless Frontier: A Report to the President*. Washington, DC: US Government Printing Office.

Carson, R. 1962. *Silent Spring*. Boston: Houghton Mifflin.

Cranor, C. 2011. *Legally Poisoned: How the Law Puts Us at Risk from Toxicants*. Cambridge, MA: Harvard University Press.

Dusenbery, M. 2018. *Doing Harm: The Truth about How Bad Medicine and Lazy Science Leave Women Dismissed, Misdiagnosed, and Sick*. New York: HarperOne.

Gaskell, A. 2019. "Technology Isn't Destroying Jobs, but Is Increasing Inequality." *Forbes* (May 3). Available online at www.forbes.com/sites/adigaskell/2019/05/03/technology-isnt-destroying-jobs-but-is-increasing-inequality/#10ecc4de5e78. Accessed February 3, 2020.

Gibbons, M. 1999. "Science's New Social Contract with Society." *Nature* 402: 81–84.

Guston, D.H. 2000. "Retiring the Social Contract for Science." *Issues in Science and Technology* 16 (4): 32–36.

Jenkins, M. 2011. *What's Gotten into Us: Staying Healthy in a Toxic World*. New York: Random House.

Johnson, P.A., T. Fitzgerald, A. Salganicoff, S.F. Wood, and J.M. Goldstein. 2014. "Sex-Specific Medical Research: Why Women's Health Can't Wait." Report of the Mary Horrigan Connors Center for Women's Health & Gender Biology, Brigham and Women's Hospital. Available online at www.brighamandwomens.org/assets/bwh/womens-health/pdfs/connorsreportfinal.pdf. Accessed January 15, 2018.

Kearns, C.E., D. Apollonio, and S.A. Glantz. 2017. "Sugar Industry Sponsorship of Germ-free Rodent Studies Linking Sucrose to Hyperlipidemia and Cancer: An Historical Analysis of Internal Documents." *PLOS Biology* 15 (11): e2003460. doi:10.1371/journal.pbio.2003460.

Kearns, C.E., L.A. Schmidt, and S.A. Glantz. 2016. "Sugar Industry and Coronary Heart Disease Research: A Historical Analysis of Internal Industry Documents." *JAMA Internal Medicine* 176 (11): 1680–1685.

Keys, A. 1971. "Sucrose in the Diet and Coronary Heart Disease." *Atherosclerosis* 14 (2): 193–202.

Keys, A. 1975. "Coronary Heart Disease – The Global Picture." *Atherosclerosis* 22 (2): 149–192.

Keys, A. (ed.). 1980. *Seven Countries: A Multivariate Analysis of Death and Coronary Heart Disease*. Cambridge, MA: Harvard University Press.

Kourany, J.A. 2010. *Philosophy of Science after Feminism*. New York: Oxford University Press.

Krishna, V.V. 2014. "Changing Social Relations between Science and Society: Contemporary Challenges." *Science, Technology, and Society* 19 (2): 133–159.

Landrigan, P.J., and M.M. Landrigan. 2018. *Children and Environmental Toxins: What Everyone Needs to Know*. New York: Oxford University Press.

Latour, B. 1987. *Science in Action*. Cambridge, MA: Harvard University Press.

Leslie, I. 2016. "The Sugar Conspiracy." *Guardian* (April 7). Available online at www.theguardian.com/society/2016/apr/07/the-sugar-conspiracy-robert-lustig-john-yudkin. Accessed February 26, 2020.

Liu, K.A., and N.A. Dipietro Mager. 2016. "Women's Involvement in Clinical Trials: Historical Perspective and Future Implications." *Pharmacy Practice* 14 (1) (March). doi:10.18549/PharmPract.2016.01.708.

Lustig, R. 2009. "Sugar: The Bitter Truth." Available online at www.youtube.com/watch?v=dBnniua6-oM.

Lustig, R. 2012. *Fat Chance: Beating the Odds Against Sugar, Processed Food, Obesity, and Disease*. New York: Penguin.

Mazure, C., and D. Jones. 2015. "Twenty Years and Still Counting: Including Women as Participants and Studying Sex and Gender in Biomedical Research." *BMC Women's Health* 15 (94). doi:10.1186/s12905-015-0251-9. Available online at www.ncbi.nlm.nih.gov/pmc/articles/PMC4624369/. Accessed July 30, 2017.

McKnight, S.A. 2005. "Francis Bacon's God." *The New Atlantis* 10 (Fall): 73–100.

McKnight, S.A. 2006. *The Religious Foundations of Francis Bacon's Thought*. Columbia: University of Missouri Press.

Moss, K.L. (ed.). 1996. *Man-Made Medicine: Women's Health, Public Policy, and Reform*. Durham, NC and London: Duke University Press.

Moss, M. 2013. *Salt Sugar Fat: How the Food Giants Hooked Us*. New York: Random House.

National Institute of Diabetes and Digestive and Kidney Diseases. 2017. "Overweight and Obesity Statistics." (August). Available online at www.niddk.nih.gov/health-information/health-statistics/overweight-obesity. Accessed January 15, 2018.

Nestle, M. 2016. "Food Industry Funding of Nutrition Research: The Relevance of History for Current Debates." *JAMA Internal Medicine* 176 (11): 1685–1686.

Nestle, M. 2018. *Unsavory Truth: How Food Companies Skew the Science of What We Eat*. New York: Basic Books.

Qureshi, Z. 2019. "Harnessing Technology for More Robust and Inclusive Growth: An Agenda for Change." Brookings Up Front Report (June 3). The Brookings Institution. Available online at www.brookings.edu/blog/up-front/2019/06/03/harnessing-technology-for-more-robust-and-inclusive-growth-an-agenda-for-change/. Accessed February 3, 2020.

Rohe, W. 2017. "The Contract between Society and Science: Changes and Challenges." *Social Research: An International Quarterly* 84 (3): 739–757.

Rosser, S. 1994. *Women's Health – Missing from U.S. Medicine*. Bloomington and Indianapolis: Indiana University Press.

Sarewitz, D. 2016. "Saving Science." *The New Atlantis* 49: 4–40.

Sarewitz, D., G. Foladori, N. Invernizzi, and M.S. Garfinkel. 2004. "Science Policy in Its Social Context." *Philosophy Today*, Supplement, 67–83.

Schiebinger, L. 1999. *Has Feminism Changed Science?* Cambridge, MA: Harvard University Press.

Schiebinger, L. *et al.* n.d. *Gendered Innovations in Science, Health and Medicine, Engineering, and Environment Project.* Available online at http://genderedinnovations.stanford.edu/methods/health.html. Accessed January 15, 2018.

Siegel, R.L., K.D. Miller, and A. Jemal. 2020. "Cancer Statistics, 2020." *CA: A Cancer Journal for Clinicians* 70 (1): 7–30.

Stevens, E.S. 2002. *Green Plastics: An Introduction to the New Science of Biodegradable Plastics.* Princeton, NJ: Princeton University Press.

Thornton, J. 2000. *Pandora's Poison: Chlorine, Health, and a New Environmental Strategy.* Cambridge, MA: MIT Press.

Viergever, R.F. 2013. "The Mismatch between the Health Research and Development (R&D) that is Needed and the R&D that is Undertaken: An Overview of the Problem, the Causes, and Solutions." *Global Health Action* 6 (1). doi:10.3402/gha.v6i0.22450.

Virani, S.S., A. Alonso, E.J. Benjamin, M.S. Bittencourt, C.W. Callaway, A.P. Carson, et al. 2020. "Heart Disease and Stroke Statistics – 2020 Update: A Report from the American Heart Association." *Circulation* 141: e1–e458. doi:10.1161/CIR.0000000000000757.

Watts, G. 2013. "Sugar and the Heart: Old Ideas Revisited." *BMJ* 346: 16–18.

Weisman, C.S., and S.D. Cassard. 1994. "Health Consequences of Exclusion or Underrepresentation of Women in Clinical Studies (I)." In *Women and Health Research*, edited by A.C. Mastroianni, R. Faden, and D. Federman, vol. 2, pp. 35–40. Washington, DC: National Academy Press.

White, H.B. 1968. *Peace among the Willows: The Political Philosophy of Francis Bacon.* The Hague: Martinus Nijhoff.

Woodhouse, E.J. 2003. "Change of State? The Greening of Chemistry." In *Synthetic Planet: Chemical Politics and the Hazards of Modern Life*, edited by M.J. Casper, pp. 177–193. New York: Routledge.

Woodhouse, E.J. 2006. "Nanoscience, Green Chemistry, and the Privileged Position of Science." In *The New Political Sociology of Science: Institutions, Networks, and Power*, edited by S. Frickel and K. Moore, pp. 148–181. Madison: University of Wisconsin Press.

Yudkin, J. 1957. "Diet and Coronary Thrombosis: Hypothesis and Fact." *Lancet* 270 (6987): 155–162.

Yudkin, J. 1964. "Patterns and Trends in Carbohydrate Consumption and their Relation to Disease." *Proceedings of the Nutrition Society* 23 (2): 149–162.

Yudkin, J. 1972. *Pure, White, and Deadly: The Problem of Sugar.* London: Davis-Poynter Ltd.

Part II
Democracy and Citizen Participation in Science

6 Which Science, Which Democracy, and Which Freedom?

Hans Radder

DEPARTMENT OF PHILOSOPHY, VU UNIVERSITY AMSTERDAM

6.1 Introduction

The three terms in the title of this volume refer to complex and contested notions. An adequate discussion of how science, freedom, and democracy are, or should be, related requires a substantive explanation of these notions. Therefore, the three main sections of this chapter – on science, democracy, and freedom – first describe these notions in general terms and discuss several of their relevant features. After these explanatory discussions, each section focuses on more specific aspects and provides examples of actual or desirable connections between science, democracy, and freedom.

Concerning science, I argue for the societal relevance and public interest of basic research, the specificity of the human sciences, and the significance of education for citizenship. The central elements of democracy are taken to be voting, deliberation, the legally supported separation of the executive, legislative, and judiciary power of the state, and an appropriate and inclusive suffrage. The relation between science and democracy is addressed with the help of a brief discussion of the democratic substance of the conception and implementation of the Dutch National Research Agenda. Finally, the notion of freedom in individualistic political theories is held to be inadequate. Instead, a more convincing account of freedom proves to be included in the proposed account of democracy. Therefore, I have made a small but meaningful change in the order of the three basic notions: "democracy" precedes "freedom". From this perspective, I discuss the social, legal, and practical meaning and implications of academic freedom. A crucial point is that this freedom should include the freedom to pursue basic science. For economic and political reasons, basic science is currently under strong pressure, and thus its public interest deserves to be explicitly defended.

In sum, the chapter starts from a general account of the notions of science, democracy, and freedom, and then discusses and illustrates several implications of this account for the relationship between science, on the one hand, and democracy and freedom, on the other. Given the complexity of these notions and the limited space available, it will be obvious that this account has to be presented in a summary fashion. Although some of the specific points are developed in some detail, the other points draw on and summarize a substantial body of literature, both by myself and by many other authors.

6.2 Which Science?

The notion of science used in this chapter is the broad European notion. It qualifies as scientific all disciplines practiced in universities and other academic institutes, a classification that is widely employed in many countries and languages. Thus, there is a diversity of sciences, including formal, physical, biomedical, engineering, cognitive, social, and human sciences. As to the latter, I prefer the term "human sciences" over "humanities". The human sciences address the nature, the histories and societies, and the possible futures of humans. In particular, they see (or should see) human beings as natural creatures who *interpret and understand* themselves and their environment in ways that do not apply to non-human entities. Classifying the human sciences as scientific acknowledges them as fully legitimate members of the family of the sciences. As is the case in all families, the members of the scientific family are both similar and distinct.

The aims of science cannot be limited to the generation of knowledge (let alone theoretical knowledge), as is done in the still frequently taken-for-granted science-as-knowledge views. Real sciences include a variety of practices whose primary aim is not the acquisition of knowledge but, for instance, the experimental creation of stable phenomena, the construction of feasible computational methods, or the design and building of instruments or other devices needed for doing empirical research. Of course, these practices make use of knowledge, but generating (theoretical) knowledge is not their primary aim.

Related to this is the fact that science and technology are often interconnected, both conceptually and empirically. Acknowledging this fact is crucial for an adequate understanding of both science and technology. The point is especially relevant for philosophical approaches (like the ones in this book) that aim to contribute to democratic debates on the appropriate role of science in society. After all, the public experiences the impact of science to a large extent, or even primarily, through its contributions to technology. Acknowledging that science and technology are often intimately related does not imply that they are identical or basically similar, as is claimed by those authors who advocate a strong notion of technoscience. For this reason, it makes full sense to distinguish between basic and application-oriented science. In contrast to many authors, I define basic science in terms of generality and unpredictability, and *not* in terms of autonomy, neutrality, or curiosity (Radder 2019, 218–224). In the last section I discuss this subject in more detail.

Many recent studies have focused on the role of values in science. That values play an important role has been widely accepted. Debate has continued on the nature and significance of these values (for instance, by differentiating between epistemic and non-epistemic or between cognitive and social values) and on the question of the justifiability of the uses of such values in various scientific practices (Lacey 1999; Douglas 2009; Elliott 2017). In this chapter, I will not address this subject in a systematic fashion. I mention it because a discussion of the relation between science, democracy, and freedom will unavoidably include issues concerning the role of values in science.

Finally, the discussion of the relations between science, democracy, and freedom should not be limited to research but also address education. Primarily, this education takes place at institutes of higher education. In addition, a variety of other schools and schooling institutes may contribute to this purpose. Being educated on the important issues of our current societies and acquiring the skills of critical reflection on these issues is crucial for democratic citizenship. That science and technology, which have a significant impact on the ways in which people live, or would like to live, their lives, constitute one such issue should be obvious. Therefore, teaching for citizenship cannot be limited to the traditional humanities (in German: *Geisteswissenschaften*), which often fail to acknowledge the important role of science and technology (see, for instance, Nussbaum 2010). This kind of education should include systematic and critical reflection on the relations between science, technology, and society (Radder 2019, 232–235).

I started this section by emphasizing the diversity of the sciences and, in particular, by defending an important distinction between the human sciences and the other academic disciplines. To repeat, the human sciences see (or should see) human beings as natural creatures who *interpret and understand* themselves and their environment in ways that do not apply to non-human entities.[1] In the second part of this section, I provide a brief example of the relevance of this point for the subject of science and democracy.

In their research, human scientists not only employ their own interpretations, they will also be confronted with the interpretations of their research subjects. Anthony Giddens (1984, 281–288) explains the significance of this so-called double hermeneutic for interpretative social science and describes it as

> the intersection of two frames of meaning as a logically necessary part of social science, the meaningful world as constituted by lay actors and the metalanguages invented by social scientists; there is a constant "slippage" from one to the other involved in the practice of the social sciences.
>
> (Giddens 1984, 374)

This "logical necessity" implies that the voice of the research subjects should be welcomed and seriously taken into account. Taking seriously the interpretations of the research subjects means acknowledging the significance of a type of democratic input into the human sciences.[2]

I recently came across an illuminating example of the practical significance of the double hermeneutic (see Smolka 2019, 106–108). Psychologists standardly use the notion of emotion as a momentary state, and distinguish between negative and positive emotions. For instance, being glad is a positive emotion, being sad a negative one. However, in neuroscientific brain-imaging experiments on the possible impact of meditation on the experience of emotions, one group of research subjects, the expert meditators, explicitly included their own interpretation in classifying an emotion as negative or positive. In particular, if they are sad because someone is suffering but feel compassion with this person, they see this emotion as positive. Thus, here is a clear case of diverging concepts of emotion between the research subjects and the neuroscientists. In this case, the concepts of the research subjects need to be taken seriously, and there is no reason to see the scientific concepts as being a priori superior.

The significance of the double hermeneutic is not limited to experiments with self-reporting, like the one above. More generally, there may be differing, or even opposing, epistemic interpretations and normative assessments of what is at stake in experiments and observations or in their theoretical interpretations in the human sciences.[3] Therefore, the double hermeneutic is not merely of epistemic significance and certainly not merely a problem of potential bias that should be solved or avoided in order to rescue the objectivity of the human sciences (as many proponents of an exclusively experimental approach to the human sciences think). Although a serious consideration of the double hermeneutic does not give you a developed citizen science, it involves a specific openness of the human sciences to the lifeworld of the people. At stake from a normative perspective is the fundamental question of who is entitled to define the nature of human beings and the significance of their cultures and societies: the scientists, the involved research subjects, or both? My basic point is that neither the scientific nor the day-to-day interpretations should be taken as a priori preferable or correct. Instead, settling particular issues needs to be done on a case-by-case basis, by explicitly considering their epistemic, their social, and their normative dimensions.

6.3 Which Democracy?

My starting point in discussing the notion of democracy is a widely shared view. It defines the aim (or the value) of a democracy as a process of giving the people of a particular country or region an effective say in deciding on all matters that do or may affect them in significant ways. The following five points briefly explain how this general definition can be further developed (for much more detail, see Radder 2019, chapter 6).

First of all, procedures of voting, which guarantee the people a say on which visions, policies, and implementations will be endorsed and employed in the governance of a country or region, constitute an essential feature of any democracy. A common but stronger formulation (discussed in Kitcher 2006, 1212) is that democracy implies "having individual control" over the conditions that affect one's life. This alternative formulation is based on the notion of individual autonomy. Although it may be quite common, it is less appropriate. The reason is that having an effective say in a *collective* process does not, or not always, imply being in *individual* control. I will come back to this point in the discussion of the notion of freedom in the next section. A premise of democratic governance is that most of the time most people possess, or are able to acquire, the basic capabilities required for democratic involvement.[4] A somewhat weaker but more plausible claim is that working towards the realization of this premise is a crucial and worthwhile duty that deserves and requires an enduring commitment.

Deliberation constitutes a second important feature of a democracy. Because many visions, policies, and implementations are complex and have far-reaching consequences, high-quality deliberation requires knowledge and comprehension; and because the people involved cannot be supposed to agree on all relevant matters, a real democracy should provide concretely implemented opportunities for critical debate and citizen participation. This also presupposes the basic right of the freedom of speech (or, more generally, expression) and the support of independent public media. That is to say, a democracy cultivates institutionalized learning processes (including insights from a diversity of academic disciplines) to maintain or increase the quality of deliberation and policy making. These rights should not merely be proclaimed in an abstract way, but they must be guaranteed through a legally and institutionally supported public sphere. As Nancy Fraser (2014) argues, this public sphere should be *actually*, and not just formally, *open to everyone*; it should be *critical*, which requires examining the deeper roots (and not merely the surface phenomena) of relevant issues; finally, it should be *effective*, in the sense that its legitimate results have a real impact on actual policies, regulations, or laws. The significance of this second feature of democracy, deliberation, implies that elections during a state of emergency (in which these rights are strongly curtailed or even fully suspended) are invalid. A case in point are the Turkish elections for presidency and parliament held in June 2018 during an emergency state that had lasted for almost two years.

Third, the modern democratic state should be a constitutional democracy. Such a democracy is supported by a constitution and related laws concerning basic human rights and by the separation of the executive, legislative, and judiciary power of the state. An important normative aim of the constitution and the legal system is to protect the people as a whole against ill-considered policies resulting from the folly of the day and, in particular, to protect minorities against a potential tyranny of the majority.[5]

Related to the latter is a fourth important feature of democracy. It concerns the constitution of "the people", the demos. In a traditional formulation, the core idea of democracy is "government of the people, by the people, for the people" (Mounk 2018, 56). The second and third points are crucial. Democracy means the rule of the *entire* demos. It requires that a democratic government cares for the interests of all the people affected by its politics, not merely for those of its own supporters.

A fifth and final point derives from the related question: who constitutes the electorate? Who has the right to vote? A principled answer is that the electorate should consist of all the adult, legal, and long-term residents who are living in a certain country or region. The reason is that they are the ones who are affected by the visions, policies, and implementations of the government. Thus, democracy requires universal suffrage of the electorate. Clearly, the electorate does not fully coincide with the demos, since the latter includes young children and minors, short-term residents, and people who lack a legal status. But although there are good reasons why these groups are not allowed to vote, they should be taken into account in democratic politics. First, because a democratic society has a duty of caring for their interests; second, because they may, and regularly do, make contributions to a flourishing democratic culture through their actions. See, for instance, the recent actions by school children and minors in raising awareness of the issues of climate change.

In sum, the *value* of democracy includes these five core features: voting, deliberation, a constitution, an inclusive demos, and an appropriate and inclusive electorate. As is· generally true, values guide practices in certain directions (rather than others), but they do not determine the actual forms these practices may take. The latter requires additional interpretation and contextualization. Thus, actual voting procedures, forms of deliberation, legal systems, and definitions of the people and the electorate are not fully determined by the general value of democracy. For instance, although voting is an essential feature of democracies, its particular procedures should not be confused with the general notion of democracy as explained by its five core features. Particular voting procedures are pragmatic tools for decision-making processes. Put in philosophical terms (Radder 2006, chapters 8–11), the nonlocal meaning of the value of democracy is not identical with and exhausted by the actual voting procedures through which we attempt to realize it. Rejecting this identity is important, for three reasons. It urges us to acknowledge the question of how taken-for-granted procedures can be improved and made more democratic as a central challenge to democratic societies; furthermore, it encourages the exploration of different kinds of pragmatic democratic procedures (including well-thought-out and well-prepared referenda); finally, it underlines the crucial importance of creating and fostering a socio-cultural climate that promotes and encourages democratic engagement concerning all kinds of relevant public issues.[6]

The two following examples illustrate these points. The first is about voting procedures. It is fictive but cases like this one happen all the time. Suppose a government is based on the results of an election with a turn-out of 70% of the formal electorate. Suppose also that the government itself consists of representatives of three political parties, one having ten votes in the Cabinet meetings, while each of the other two has five votes. Under these conditions, making compromises will be a frequent feature of the decisions made by this government. Thus, it will regularly happen that particular decisions are taken that are abided (but not really supported) by the two minority parties. In this case, these decisions are actually supported by 50% of the Cabinet members. Consequently, the decisions are based on no more that 35% of the full electorate. Put differently, in actual political practice many decisions are taken that are neither the will of the entire people, nor of the voters, nor even of the majority of the voters.[7] Frequently occurring examples like this teach an important lesson, with theoretical and practical consequences. They imply that the notion of "the will of the people" is questionable and misleading. Theoretically, one might question whether it makes sense to assign some kind of "aggregated will" to collectives. More important is that in practical politics the phrase is frequently abused to legitimate decisions that are, arguably, not supported by the entire people. Therefore, in theoretical debates on democracy and in day-to-day practical politics the notion of the will of the people had better be left out completely.

The second example concerns the constitution of the electorate. The common view is that, during a long historical process, membership of the electorate has been stepwise expanded to include the elite and the masses, the rich and the poor, the male and the female, the white and the colored. In mature democracies, this process is taken to have reached its ideal final state of universal suffrage. "In a democracy … all citizens get one vote without regard to the color of their skin or the station of their ancestors" (Mounk 2018, 130).

But is this really the case? The widely shared view of democracy, with which I started this section, implies that the electorate should consist of all the adult, legal, and long-term residents who are living in a certain country or region. In practice, however, in quite a few countries participation in (national) elections is not based on this kind of residency but on nationality or formal citizenship. This implies that, on the one hand, legal, long-term residents who (for some reason or other) do not have this nationality or citizenship are unjustly excluded from democratic rights. On the other hand, it inappropriately keeps granting these rights to emigrants who have been living abroad for many years, as long as they have kept the relevant nationality or citizenship. Depending on the specific countries, the two groups may have considerable sizes.[8] These facts further detract from the representativeness of actual decision-making procedures, and so provide another reason to avoid the idea of the alleged will of the people. Instead, in a real democracy membership of the electorate needs to be based on residency, and not on nationality. In his recent book, Yasha Mounk rightly argues for the importance of an "inclusive patriotism" (2018, 207–215). A democracy requires the involvement of its citizens in the fortunes of the country or region in which they live. This involvement needs to be appropriate (thus, excluding those citizens who live abroad and are not, or hardly, affected by the relevant politics) and inclusive (thus, open to all the adult, legal, and long-term residents). Such an appropriate and inclusive suffrage would bring the realization of an inclusive patriotism one step closer.

Let us now turn to the subject of science and democracy. This is a complex subject that can be addressed from a variety of perspectives.[9] In the second part of this section, I review a specific case of public participation in science policy and evaluate its democratic substance on the basis of the five core features of democracies explained in its first part. The case is the recent development of the Dutch National Research Agenda (DNRA). In Radder (2019, 244–251), I have provided a detailed account of this case, including the views of a variety of other commentators. Here, I first provide a summary and update of the implementation of the DNRA and then assess its democratic quality.

The aim of the DNRA was to consult the Dutch citizens about the question of which topics should be included in the research agenda of the Dutch scientists. It implied a stronger direction of the sciences toward central societal issues. The DNRA is claimed to be a unique form of citizen science, the results of which should show that science4competitiveness. The implementation of the DNRA, directed by a so-called Knowledge Coalition, started in 2015. The DNRA website summarizes the process as follows:

> The Dutch approach to developing a national science agenda has been genuinely unique, in the way that the Knowledge Coalition invited the general public in April 2015 to submit questions about science. This resulted in 11,700 questions submitted by the general public, academic institutions, the business community and civil society organisations. Five academic juries were appointed by the Knowledge Coalition to cluster and assess the questions. This was followed by three conferences in June 2015 on "science4science", "science4competitiveness" and "science4society". Their purpose was to bring order to the questions, and to further aggregate the questions where possible, based on these three perspectives. This process ultimately led to the 140 overarching scientific questions and 16 example routes.
>
> (Dutch National Research Agenda 2015)[10]

In agreement with the approach in this chapter, the notion of science is used in its broad European sense and the DNRA does not presuppose a strong contrast between science and technology.

To make the huge set of questions practically manageable required substantive processing procedures. Most of this incisive processing was done by the mentioned academic juries. It resulted in summarizing the 11,700 questions into 140 overarching questions. Based on the keywords added by the applicants to their questions, Malou van Hintum (2015, 23–24) provides the following rough but informative ranking of the topics of the most frequently posed 25 questions: health (480); energy (370); the brain (280); building (277); sustainability (270); education (260); the economy (242); development (212); behavior (192); chemistry (185); children (171); innovation (165); cancer (162); sustainable (161); security (156); prevention (154); language (147); climate change (146); data (146); sport (145); history (141); well-being (141); life (138); culture (135); society (128). In a first round of funding, the Dutch government made available a budget of 61 million euros. Teams consisting of scientists and commercial or societal partners have submitted research projects that address the overarching questions. These projects have been assessed by means of the usual science policy procedures. In June 2019, the results of this funding process have been made public: out of 323 proposals, 17 research projects will be actually funded.

Can this DNRA be seen as a successful form of citizen science (or citizen science policy) and, more broadly, as an exemplar of democratizing science? Unfortunately, this is not the case. The reason is that both its practical implementation and its conceptualization of the relation between science and society suffer from a variety of significant democratic deficiencies.

1 First, there is the composition of the Knowledge Coalition, and the Steering Group appointed by it, which is quite one-sided and biased toward economic interests. The Coalition consists of representatives of employers' associations and of established policy organizations in the fields of higher education, technology, and medicine. Labor unions, NGOs, and grassroots collectives are conspicuously absent.

2 Second, the claim that "all stakeholders were involved" (De Graaf, Rinnooy Kan, and Molenaar 2017, 236) is not correct. At about the same time as the DNRA, a variety of academic grassroots organizations launched strong and broadly supported criticisms of the commodification, hierarchization, and bureaucratization of Dutch science and university policies.[11] These stakeholders have not been involved at all.

3 Third, there are two features of the practical implementation of the DNRA that significantly detract from its alleged bottom-up character. The first is that a substantial number of the questions (in some areas even half of them) have been submitted by professional scientists, either individually or as a group (Van Hintum 2015, 24–35). The second problem is that the process of clustering the 11,700 questions into the 140 overarching ones, which has been carried out exclusively by academics, was by no means a neutral procedure. The following example shows what gets lost in this process. One question reads: "How can we increase the inclusion, resilience and talents of young people with (mental, cognitive, physical) developmental problems and retardations?" According to one of the scientific jury members, the core of this question can be rephrased as: "How can we have children grow up safely?" (Van Hintum 2015, 51–52). But of course the original question is not limited to questions of safety. And while the latter question could be used to promote technocratic research about security policies, the former question could entail a rather different set of research projects.

4   The fourth problem concerns the way the issues have been split up in terms of the questionable distinction between science for science, science for society, and science for competitiveness, a distinction that is also used in current European science policy. Apparently, basic science is seen as "not for society", social and economic issues are taken to be neatly separable, and the economy is reduced to competitiveness.[12]

5   A fifth issue is the lack of reflexivity. The explanation of one of the questions states that the DNRA needs a more realistic view of what science can, and cannot, achieve. At the end of the day, however, this important request for reflexive understanding has hardly been honored: the final list of 140 overarching questions includes only two of such explicitly reflexive questions. One is about the normative issue of what is "good" science; the other about the possible limits of scientific knowledge (see Van Hintum 2015, 167–174).

6   A final problem concerns the funding procedures and the resulting decisions on approval and rejection of the submitted projects. The results of the first round of applications (which became available in June 2019; see NWO 2019) show the usual deficiencies and biases of this type of science funding (cf. Radder 2016, 93–101). For a start, the success rates have been extremely small: a mere 5.26% of the submitted projects have been approved. Furthermore, as can be gathered from the above-mentioned frequencies of the top 25 keywords, the full set of questions covers a remarkably broad spectrum of subjects that require both basic and application-oriented research from all academic disciplines. Beatrice de Graaf emphasizes the same point: "Among the ... queries submitted online, many asked questions having to do with the origins of mankind, with society's resilience, with identity, democratic citizenship, and the need for spirituality and religiosity. Utilitarian motives did not predominate" (De Graaf 2017, 189).

In spite of this, the funded projects show a strong bias toward the technological, physical, and biomedical sciences. These disciplines received 90.5% of the available 61 million euros; left over for the social and human sciences was a poor 9.5% (Bol 2019). Moreover, "utilitarian motives" did predominate the funding phase: only a minority of the funded projects (my estimate is 5 or 6 out of 17) can be seen as primarily consisting of basic research.

What to conclude about the democratic quality of the DNRA? I will address this question with the help of the five core features of democracy discussed in the first part of this section. The main question concerning *voting* is this: did the general public have an effective say in the decisions concerning the content of the Dutch research agenda? The answer is: only in a very limited way. In principle, every citizen had the opportunity for submitting questions. But due to the specific institutional framing of the DNRA (points (1) and (3) above) its actual implementation included significant top-down procedures and a clear bias toward economic issues.[13] Furthermore, the summarizing and clustering by scientists has led to significant displacements and distortions of the original questions posed by the public (point (3)). Finally, although the decision-making processes about the funding of the submitted research projects have not been made public, such processes often include some kind of voting procedures by scientists and policy makers. Whatever the precise details of these procedures, their outcomes show a significant mismatch between the skewed disciplinary distribution of the approved projects and the much broader disciplinary variety of the topics of the questions submitted by the public (point (6)).

In the course of developing the DNRA, a variety of forms of *deliberation* have taken place, both in academic and in popular settings. Yet, the quality of these deliberations has been substantially impaired by the conceptual framing of the presentations and debates. As I concluded in points (4) and (5), the DNRA suffered from substantial conceptual shortcomings. They concern the implied divides between science, society, and the economy and the lack of substantial reflection on what we can reasonably expect from science for the purpose of solving our societal problems.

The third core feature of democracies is the support they receive from *an independent constitution and legal system*. In the case of the DNRA, this support is rather indirect. No special laws have been passed for the purpose of creating a Dutch research agenda. At most, one could say that policy making on an important societal issue like the social relevance of science is a legally supported task of a democratically chosen government. There is one specific question, though, that is relevant in this context. Although developing a DNRA has been broadly supported, there have also been criticisms claiming that it goes against the idea of academic freedom (Van Hintum 2015, 108–109 and 135–136; Molenaar 2017, 33–36). However, this criticism (which sees *any* public direction of academic research as illegitimate) can itself be questioned. First, as we will see in the next section, in the Dutch context the precise meaning and implications of the legal notion of academic freedom are unclear and disputed; second, the DNRA does not seem to impair a basic point of academic freedom, which is that it is still the scientists themselves who decide on the choice of the appropriate methods to study the relevant questions and on the acceptability of the proposed answers to these questions.

The two final core features are the *inclusiveness of the demos* and *the appropriateness and inclusiveness of the electorate*. In the first stage of the process, the collection of the questions, all citizens of the Netherlands, even children, were allowed to participate. However, this formal openness did not lead to a balanced inclusion of the entire demos (point (3)). Allowing the submission of questions by scientists, and even groups and organizations of scientists, has given this specific part of the public an unrepresentative and hence inappropriate weight. Furthermore, the summarizing and clustering was exclusively done by the members of the scientific juries, thus excluding wider participation of, and opportunities for genuine interaction with, the broader public at this stage. Finally, point (6) makes one quite suspicious about the representativeness of the (scientific) electorate of the committees that have made the final funding decisions. Although the detailed disciplinary composition of these committees and their external reviewers has not been made public, the hypothesis that it has not been proportional to the distribution of the questions over the different academic disciplines is quite plausible.

My conclusion from this analysis is twofold. Negatively, we have to conclude that calling the DNRA a case of citizen science (or citizen science policy) is already questionable, but it is certainly not a successful example of democratizing science. Positively, this way of analyzing this attempt at democratizing science makes explicit the relevant issues that have to be taken into account and entails concrete ways of where and how possible future attempts should and could be improved.

6.4 Which Freedom?

Quite a few people employ and advocate the concept of "liberal democracy". Such views frequently assume two independent philosophical and socio-political notions, liberty and democracy. Next, the two are somehow merged into the notion of liberal democracy. The former, liberty, is often seen as the primary one and interpreted in terms of individual freedom and autonomy as the basic qualities or ends of human beings. Freedom, then, is chiefly negative freedom, the absence of interference by others. In line with this, societies are conceptualized as aggregates of individuals, each with his or her own properties and preferences. Given the primacy of individual freedom and autonomy, democracy may even be seen as a problem, namely, the problem of making practicable concessions needed to resolve the actual and potential tensions between the distinct interests of the individual members of society. From this perspective, the best solution is the one that guarantees an optimal (negative) freedom and autonomy for all citizens.

While many philosophers and politicians subscribe to such views, many others have rightly criticized these as basically mistaken. The problem is two-fold. First, individual people are not some kind of independent, atomistic entities. Their interactions with others constitute an essential part of what they are, of their beliefs and desires, their hopes and fears, their histories and futures. Second, their societies possess many supra-individual characteristics that structurally constrain the freedom of their individual members. These constraints are often persistent. Although they can be changed in principle, in practice the freedom of actual individuals is often limited in fundamental ways because of the presence of supra-individual entities that structurally shape the ways they live, and have to live, their lives.[14]

In Radder (2019, chapter 6) I have discussed these points in detail and illustrated them with the (imagined) example of a certain Mr. Freeman. Because of the 2008 financial crisis, he has become extremely critical of the global financial system. Therefore, he wants to have nothing to do anymore with banks and other big monetary organizations and has decided to do all his financial transactions in cash. Very soon, however, he is confronted with the disastrous consequences of this "free" decision: leading an ordinary life without a bank account has become virtually impossible in our current societies. Moreover, the global financial system is only one of many "large technological systems" that structurally and persistently constrain the freedom of individuals. Such systems constitute genuine interests that cannot be reduced to some aggregate of the individual interests of the current members of a society. In individualistic political theories, such interests are either ignored or not recognized at all.

For these reasons, I do not start with first positing a separate sphere of individual human freedom and autonomy, which then has to be somehow reconciled with the limitations of the socio-material life of individuals. Freedom cannot be the exclusive or primary theoretical or ontological idea of accounts of human nature and human societies. Instead, the relevant socio-political notions can be directly derived from the (partly descriptive, partly normative) comprehensive concept of democracy sketched in the preceding section. The conditions that need to be satisfied for appropriately practicing this kind of democracy include, first, the freedom of expression; second, the right to participate, individually or collectively, in a public sphere; third, the legal and institutional support of this public sphere, in such a way that it allows and advances legitimate, critical, and effective involvement in democratic decision making; fourth, this kind of involvement also presupposes a right to education as a basic human right. Therefore, terms like "illiberal democracy" or "democracy without rights" (Mounk 2018) constitute contradictions in terms.[15] To be clear, such a view of the relation between democracy and freedom does not deny that having individual rights constitutes an important regulative value. Therefore, where appropriate this view may be supplemented with a normative account of basic morality, especially regarding the important issues of justice. Thus, this approach acknowledges the significance of general moral values, but it simultaneously highlights the entanglement (rather than the separation) of these values and their socio-political interpretations and contextualizations.

As we have seen, the basic aim of a democracy is to give the people of a particular country or region an effective voice in deliberating and an effective say in deciding on all matters that do or may affect them in significant ways. The matters that affect them do not merely include the negative conditions that interfere with their interests but also the positive circumstances that allow citizens to concretely develop the capabilities they have. That is to say, a democratic society needs to advance not only the negative but also the positive freedom of its members.

In sum, this view of the relation between democracy and freedom goes against the ontological account of human beings as free and autonomous individuals; it similarly rejects the implausible account of society as an aggregate of autonomous individuals. Instead, this view includes the constitutive role of socio-material interactions and it acknowledges the empirical and normative significance of supra-individual entities. To be clear, this does not imply a fatalistic compliance with the status quo. On the contrary, it is only critical reflection on these socio-material interactions and supra-individual entities that makes possible fruitful policies and actions concerning *avoidable*, identified and unidentified, oppression of freedom (cf. Kitcher 2006, 1213–1215).

What are the implications of the preceding account of democracy and freedom for science? This question addresses a wide-ranging topic. In the context of this book, its discussion will be limited to academic freedom. This type of freedom derives from the professional status of academics and their specific capabilities. It should not be confused with the political freedom academics share with all citizens of a democratic state. Yet, the two are significantly related. Without such political freedom, exercising the right of academic freedom will be very difficult, if not impossible.

It is undeniable that science affects the lives of many people in substantial ways. Therefore, citizens of a democratic state possess the right to have an effective say concerning scientific matters. As we have seen in Section 6.2, this includes both basic and application-oriented sciences as well as their relations to technology. The crucial question is, of course, the nature and scope of this effective say. Clearly, this cannot mean that the public at large (including political or commercial parties) could simply vote on the acceptability of scientific methods and knowledge claims. Adoption of such a procedure would deny the specific value and quality of deliberation in science and its specific contributions to matters of public interest. These specificities clearly justify a measure of academic freedom in how scientific institutes organize their basic methodological and epistemic procedures.[16] This view agrees with Mark Brown's account of the politicization of science, an account that implies that "it is not necessarily undemocratic that some citizens have more persuasive power than others in public deliberation. ... As long as such hierarchies serve public interests and do not involve hidden forms of domination, they may be democratically legitimate" (Brown 2015, 19).

In a legal sense, the precise meaning of the notion of academic freedom is often unclear (Groen 2017). The Dutch Act on Higher Education and Scientific Research, for instance, does not even provide a definition. It merely states that "institutes of higher education and academic hospitals comply with academic freedom" (Overheid 2019; my translation). Therefore, its meaning needs further explanation, which may result in distinct (legal) interpretations. Thus, Joris Groen writes:

Academic freedom as a principle of governance can be defined as a principle of participation of the academician, in collaboration with fellow academics and within the limits of the law, in decision-making about the essential elements of education and academic research. ... Academic freedom as a principle of governance does not lead to a decisive vote for the individual academic professional. ... The boundary between the essential elements of education and research and administrative and financial affairs has to be determined in practice, and will always be controversial.

(Groen 2017, 25)

Some of these "essential elements", for instance, the freedom of expression or the assessment of the quality of scientific methods and knowledge, are relatively uncontroversial. Furthermore, administrative and financial affairs do influence the quality of teaching and research. Currently, this influence is pervasive and not always positive, to put it mildly (see Halffman and Radder 2015; 2017). Therefore, an essential aspect of academic freedom is the participation in critical academic debate and decision making on what does, and what does not, constitute an optimal policy for delivering high-quality research and teaching. This means, for instance, that clauses in employment contracts saying that academics should "refrain from negative publicity" do violate the legitimate academic freedom of expression.

Another aspect of the Dutch Act on Higher Education and Scientific Research does imply a legal constraint on academic freedom. This Act states that, in addition to teaching and research, the third basic task of universities is "to pass on knowledge for the benefit of society" (since 2004, this is often called "valorization"). Therefore, if ordinary citizens also have a say in what may benefit society (which is reasonable enough), their judgments will legally constrain the institutional freedom of science in choosing its research agenda.

Thus, as a "principle of governance" academic freedom is primarily an institutional right. Yet it also entails a measure of academic freedom for individual teachers and researchers. Again, however, this individual freedom is often constrained, in ways that depend on the specifics of the institutional context. As lecturers, our teaching will be part of prestructured degree programs. If I have been appointed for teaching courses in philosophy of technology, I am not free to shift suddenly to teaching completely different topics that do not fit in the plan of the degree program.[17] Furthermore, genuine universities are based on the unity of teaching and research. As a rule, academic lecturers should also be significantly involved in research in their teaching area. This entails a further constraint on the academic freedom of individuals in their choice of research topics. Moreover, in many disciplines, doing research requires the use of expensive instrumentation and extensive collaboration in teams. Also, these facts significantly constrain the freedom of individuals in their choices of research topics. This applies in particular to the physical and biomedical sciences (and to a lesser extent to some of the social sciences). A once implemented research trajectory can be modified, but only in the longer term. In the human sciences the situation is usually different, since much research consists of small-scale, individual projects. This fact explains why the issue of *individual* academic freedom comes up mostly in these sciences, whereas it is far less prominent (or hardly an issue) in other disciplines.

Christian Krijnen (2011, 2019) has developed a detailed philosophical justification of the idea of academic freedom. It is based on two central assumptions. First, the core of university education is, or should be, the close interaction between teaching and research; second, "the value of academic research should ... be understood in its general contribution to humanity" (Krijnen 2019, 2–3; cf. 2011, 36–42). My approach to academic freedom agrees with these two assumptions but deviates from some of the conclusions drawn from them. First, my account of democracy affirms the role of freedom in cultivating humanity for all humans, not just for university scientists. While Krijnen may well agree with this broader claim (see Krijnen 2011, 42), it would seem to require some further arguments, because not all education is based on a unity of teaching and research. Furthermore, Krijnen's focus is on theoretically justified beliefs, on knowledge and on truth (2011, 42–46; 2019, 2–3, 11). However, this focus on beliefs and their justification takes no account of the *inherent* connection between empirical science and technology. Even if this connection is complex (Radder 2019, chapters 1 and 2), it implies that university education is not only a matter of *intellectual* self-activity and it points to an inherent (and not merely a contingent or secondary) relation between empirical science, technology, and, hence, social usefulness. Lastly, Krijnen (2019, 11) interprets science's contribution to humanity as *Bildung*, as a "freedom qua intellectual self-determination", aimed at individual personality formation. From the perspective on the relations between science, democracy, and freedom sketched in this chapter, this individualistic interpretation in terms of *Bildung* is questionable (for more on this issue, see Radder 2019, 232–235).

Finally, what constitutes, from this perspective, the democratic right of the citizens concerning the practice of science? My answer is: the right of having science practiced in the public interest.[18] A crucial point is that this includes not only application-oriented but also basic science. The discussion of the case of the Dutch National Research Agenda in the previous section included several examples of how the public interest of application-oriented science can be assessed and possibly improved. Here my focus is on basic science, and my claim is that the promotion and support of basic science, which can be defined in terms of generality and unpredictability, is a matter of social relevance and public interest. In Radder (2019, chapter 7) I have developed this claim in detail. I cannot repeat this discussion here. What I will do is argue that basic science is under strong pressure and needs much more support than it receives at the moment. Thus, an academic freedom that respects the right of citizens on science in the public interest should include the institutional freedom to advance basic science in a substantial way. In this respect, I agree with Krijnen that this kind of research is under strong pressure and deserves to be defended.

That basic science is under pressure has been widely acknowledged. Consider the following two examples from the Netherlands. First, the Regulations for Obtaining a Doctoral Degree at Maastricht University state that each doctoral dissertation thesis should include an obligatory section on knowledge valorization. "The doctoral candidate must add a valorisation addendum of approximately 5 pages to the thesis" (Maastricht University 2018, 47). Following the official definition used by the Dutch government, valorization is "the process of creating value from knowledge, by making knowledge suitable and/ or available for social (and/or economic) use and by making it suitable for translation into competing products, services, processes and new activities" (Maastricht University 2018, 47; for some examples of valorization, see 47–48). Even if this valorization addendum does not count in the final assessment of the thesis, its obligatory nature and its size tell us that basic science, which by definition does not aim at this kind of valorization, is not good enough.

Second, consider one of the big funding schemes of the Netherlands Organisation for Scientific Research (NWO), the Innovational Research Incentives Scheme (also called, veni-vidi-vici). Applications for funding are assessed on the basis of three criteria: the quality of the researcher (40%), the quality, innovative character, and academic impact of the proposed research (40%), and the prospects of knowledge utilization (20%). Added to the specification of the latter is the following:

The assessment committee assesses:

- whether the applicant has given a realistic description of the potential for knowledge utilisation; …
- if there is no potential for knowledge utilisation: the applicant's arguments as to why the proposed research does not lend itself to knowledge utilisation.

(NWO 2020, 16)

How to interpret these statements? The obvious explanation by the applicant should be: "my plan is to do basic research, so I do not intend to look at (immediate) applications". If basic research were fully legitimate, this answer should suffice. However, this explanation cannot be what NWO has in mind, since it appeals to a definition and not to an argument that could be assessed by the relevant committee. Therefore, the only legitimate answer for the applicant seems to be: "I have looked everywhere and I could not find a single possible application". It will be clear that this answer makes the applicant extremely vulnerable to the counterargument that he/she has not searched closely enough and therefore not "given a realistic description of the potential for knowledge utilisation". Moreover, for a clever committee member it will not be too difficult to concoct some potential application. What this shows is that the norm is that there should be knowledge utilization (or valorization). If not, you need a special but easily defeasible explanation. Given the current science-policy climate and the extremely tough competition for funding, submitting an "abnormal" basic research project will definitely lower your chances of being funded.

These two examples have been taken from the Netherlands. As we have seen in the preceding section, the funding decisions of the DNRA display the same pattern. More generally, in many countries basic research, in particular in the human sciences, is under severe pressure and sometimes even marginalized (see for instance the country reports about Japan, Brazil, and Hungary in Halffman and Radder 2017). In this context, a crucial task is to defend the *societal* value of basic science, and not merely interpret it in terms of individual curiosity.

Basic science provides epistemic and technical resources that are optimally multipurpose and open ended. Therefore, it offers the best possibilities for coping with future complexity and uncertainty. It constitutes an indispensable infrastructure that helps society to cope not just with current complex problems but also with hard-to-anticipate future complexities. In as far as science is useful for the purpose of anticipating future complexity and uncertainty, it is basic science rather than the much more specific application-oriented disciplines.[19] Therefore, a comprehensive, critical philosophy that is not limited to the particular problems of specific target groups but also includes the interests of those future generations that will be affected by our current policies needs to defend the public interest of basic research (see Radder 2019, chapters 6 and 7).

A good enough reason for promoting basic science from a societal perspective is that, not always but frequently enough, the results of this kind of research prove to constitute a significant and indispensable *component* of processes of invention and innovation. For this reason, politicians, policy makers, and university administrators should stop paying mere lip service to the value of basic research and instead support it through concrete science policies. For instance, by earmarking a substantial part of the budget of funding agencies (say, 50%) for projects in basic research, where applicants do not have to answer unanswerable questions, as they have to do in the NWO case. To conclude: unlike the advocates of "autonomous" science I do not object to valorization or knowledge utilization as such. What I argue for is, first, the public interest of application-oriented science. Second, this argument is expanded to include basic research, because promoting and supporting this kind of research also constitutes a legitimate public interest.

Acknowledgements

For helpful comments on an earlier version of this chapter I thank Ruud Abma, Mark Brown, Christian Krijnen, and the editors and reviewer of this volume.

Notes

1 This view is clearly at odds with naturalistic philosophies of science. For some criticisms of these philosophies, see Radder (1996, 175–183).
2 Standpoint theory makes the same point more sharply, tending to privilege situated lifeworld knowledge over generalized scientific knowledge. As Sandra Harding (2009, 194) states: "In hierarchically organized societies, the daily activities and experiences of oppressed groups enable insights about how the society functions that are not available – or at least not easily available – from the perspective of dominant-group activity". A different, but equally strong position is taken by Julian Reiss (2019) in his critique of (especially, social-scientific) "epistocracy".
3 See the discussion of the foot-in-the-door experiment and its connection to the replication crisis in social science in Radder (2019, 228–231). See also Feenberg (1995, chapter 5), for a case in which AIDS patients challenged the established separation between methodological cure and ethical care.
4 See I.F. Stone's account of the idea of democracy in ancient Athens (Stone 1989, chapter 4). See also Bregman (2019, chapter 15); the title of the latter book can be translated as *Most People are Alright*.
5 This plea for a broad and strong democracy does not imply that revolution is necessarily illegitimate. The above analysis bears upon democratic societies. If a society is undemocratic (for instance, because of the use of pervasive structural violence sanctioned by politics or law), revolutionary change may be legitimate.
6 The analysis and evaluation of the current situation in Mounk (2018) focuses almost exclusively on representative democracies at the national level. This leaves out the many activities of citizens in "sub-political" contexts that display and advance a democratic attitude and local democratic practices (see, e.g., Nahuis and Van Lente 2008; Nash 2014; Brown 2015). Examples are the local actions for a say in environmental issues, for concrete empowerment of women, for more democratic universities, and so on.
7 The above example concerns a particular type of democratic governance, but similar examples can be easily found, for instance in democracies based on a district system, in which the winners take it all. For instance, in the recent elections in the UK in December 2019, the Conservative Party acquired 56.2% of the seats in the House of Commons on the basis of 43.6% of the votes.
8 A telling illustration is the case of Hungary, where the sizeable "Hungarian" minorities in Romania and Slovakia are openly pampered by the current Hungarian government for their electoral support.
9 See, for instance, Brown (2009) and, of course, the other contributions to this volume.
10 In 2016 nine additional routes have been added. The process is described in detail in Van Hintum (2015, part 1) and De Graaf, Rinnooy Kan, and Molenaar (2017, 225–238); quotations from the former have been translated by me.
11 See Halffman and Radder (2015); the Dutch-language version of this *Academic Manifesto* was published in 2013.
12 For extensive criticisms of these assumptions, see Radder (2019, chapters 3–5 and 7).
13 For an insightful analysis of the notion of framing and its implied features of inclusion and exclusion in connecting science, policy, and society, see Halffman (2019).
14 These arguments are even more forceful against the stronger notion of autonomy, in the sense of individuals who are claimed to "make their own laws".
15 Note also that having (legal or moral) rights does not automatically imply being free to exercise these rights. The reason is that the exercise of rights is often

susceptible to trade-offs or dependent on constraining conditions. For example, in a democracy all members of the electorate possess the right to have an effective say in decisions on relevant public matters. Yet, the people who voted for the Liberal Democratic Party in the UK election in December 2019 proved not to have the freedom to realize a really effective say: although their share in the total number of votes increased from 4.2% to 11.6%, this gain resulted even in a loss of one of their party's seats in the House of Commons (from 12 to 11).

16 Still, the specific substance of this freedom requires further debate, as we have seen in the discussion of the double hermeneutic in Section 6.2.

17 But note that, in a democratic university, I should have had an effective voice and say in the decisions that resulted in the current degree program.

18 As I argue in detail in Radder (2019, chapter 6), assigning a public interest to a certain state of affairs essentially requires democratic support.

19 It is true that basic research projects may arise as a consequence of questions that originated in application-oriented science (see Carrier 2004, 287–291); more specifically, it may also be the case that application-oriented research leads to unexpected empirical results, as Bedessem and Ruphy (2019) argue. These points do not invalidate the above argument, however. First, there is the empirical question of how often this happens. Second, and more importantly, the legitimacy and fundability of the subsequent projects in basic research requires an argument like the one presented above.

References

Bedessem, B., and S. Ruphy. 2019. "Scientific Autonomy and the Unpredictability of Scientific Inquiry: The Unexpected Might Not be Where You would Expect." *Studies in History and Philosophy of Science* 73: 1–7.

Bol, T. 2019. "Verdeling onderzoeksgeld moet eerlijker en directer." *De Volkskrant* (June 24).

Bregman, R. 2019. *De meeste mensen deugen: Een nieuwe geschiedenis van de mens.* Amsterdam: De Correspondent. [English translation: Bregman, R. 2020. *Human Kind: A Hopeful History.* London: Bloomsbury.]

Brown, M.B. 2009. *Science in Democracy: Expertise, Institutions, and Representation.* Cambridge, MA: MIT Press.

Brown, M.B. 2015. "Politicizing Science: Conceptions of Politics in Science and Technology Studies." *Social Studies of Science* 45 (1): 3–30.

Carrier, M. 2004. "Knowledge and Control: On the Bearings of Epistemic Values in Applied Science." In *Science, Values, and Objectivity*, edited by P. Machamer and G. Wolters, pp. 275–293. Pittsburgh, PA: University of Pittsburgh Press.

De Graaf, B. 2017. "Free-range Poultry Holdings: Living the Academic Life in a Context of Normative Uncertainty." In *The Dutch National Research Agenda in Perspective: A Reflection on Research and Science Policy in Practice*, edited by B. de Graaf, A. Rinnooy Kan, and H. Molenaar, pp. 181–191. Amsterdam: Amsterdam University Press.

De Graaf, B., A. Rinnooy Kan, and H. Molenaar (eds.). 2017. *The Dutch National Research Agenda in Perspective: A Reflection on Research and Science Policy in Practice.* Amsterdam: Amsterdam University Press.

Douglas, H.E. 2009. *Science, Policy, and the Value-free Ideal.* Pittsburgh, PA: University of Pittsburgh Press.

Dutch National Research Agenda. 2015. "Approach." Available online at https://wetenschapsagenda.nl/approach/?lang=en.

Elliott, K.C. 2017. *A Tapestry of Values: An Introduction to Values in Science.* New York: Oxford University Press.

Feenberg, A. 1995. *Alternative Modernity: The Technical Turn in Philosophy and Social Theory.* Berkeley: University of California Press.

Fraser, N. 2014. "Transnationalizing the Public Sphere: On the Legitimacy and Efficacy of Public Opinion in a Post-Westphalian World." In *Transnationalizing the Public Sphere,* edited by K. Nash, pp. 8–42. Cambridge: Polity.

Giddens, A. 1984. *The Constitution of Society: Outline of the Theory of Structuration.* Cambridge: Polity Press.

Groen, J.R. 2017. "Academische vrijheid: Een juridische verkenning." PhD diss., Erasmus University, Rotterdam. Available online at https://repub.eur.nl/pub/95506/.

Halffman, W. 2019. "Frames: Beyond Facts versus Values." In *Environmental Expertise: Connecting Science, Policy, and Society,* edited by E. Turnhout, W. Tuinstra, and W. Halffman, pp. 36–57. Cambridge: Cambridge University Press.

Halffman, W., and H. Radder. 2015. "The Academic Manifesto: From an Occupied to a Public University." *Minerva: A Review of Science, Learning and Policy* 53 (2): 165–187.

Halffman, W., and H. Radder (eds.). 2017. International Responses to the Academic Manifesto: Reports from 14 Countries. Special report, *Social Epistemology Review and Reply Collective,* 1–76. Available online at https://social-epistemology.com/2017/07/13/international-responses-to-the-academic-manifesto-reports-from-14-countries-willem-halffman-and-hans-radder/.

Harding, S. 2009. "Standpoint Theories: Productively Controversial." *Hypatia: A Journal of Feminist Philosophy* 24 (4): 192–200.

Kitcher, P. 2006. "Public Knowledge and the Difficulties of Democracy." *Social Research* 73 (4): 1205–1224.

Krijnen, C. 2011. "Die Idee der Universität und ihre Aktualität." In *Wahrheit oder Gewinn? Über die Ökonomisierung von Universität und Wissenschaft,* edited by C. Krijnen, C. Lorenz, and J. Umlauf, pp. 25–51. Würzburg: Königshausen & Neumann.

Krijnen, C. 2019. "Academic Freedom as Radical Freedom." Forthcoming in *Handbook on Academic Freedom,* edited by M. Olssen, R. Watermeyer, and R. Raaper. Cheltenham: Edward Elgar.

Lacey, H. 1999. *Is Science Value Free? Values and Scientific Understanding.* London: Routledge.

Maastricht University. 2018. "Regulations for Obtaining the Doctoral Degree." Available online at www.maastrichtuniversity.nl/support/phds.

Molenaar, H. 2017. "A Plurality of Voices: The Dutch National Research Agenda in Dispute." In *The Dutch National Research Agenda in Perspective: A Reflection on Research and Science Policy in Practice,* edited by B. de Graaf, A. Rinnooy Kan, and H. Molenaar, pp. 31–45. Amsterdam: Amsterdam University Press.

Mounk, Y. 2018. *The People vs. Democracy: Why our Freedom is in Danger and How to Save It.* Cambridge, MA: Harvard University Press.

Nahuis, R., and H. van Lente. 2008. "Where are the Politics? Perspectives on Democracy and Technology." *Science, Technology, & Human Values* 33 (5): 559–581.

Nash, K. 2014. "Towards Transnational Democratization?" In *Transnationalizing the Public Sphere*, edited by K. Nash, pp. 60–78. Cambridge: Polity Press.

Nussbaum, M. 2010. *Not for Profit: Why Democracy Needs the Humanities*. Princeton, NJ: Princeton University Press.

NWO. 2019. "NWA-ORC Awarded Grants Round 2018." Available online at www.nwo.nl/en/research-and-results/programmes/nwo/dutch-national-research-agenda—research-along-routes-by-consortia-orc/awarded-grants-2018.html.

NWO. 2020. "Innovational Research Incentives Scheme – Call for Proposals 2020". Available online at www.nwo.nl/en/funding/our-funding-instruments/nwo/innovational-research-incentives-scheme/veni/ew/innovational-research-incentives-scheme-veni-enw.html.

Overheid. 2019. "Wet op het hoger onderwijs en wetenschappelijk onderzoek." Available online at https://wetten.overheid.nl/BWBR0005682/2019-02-01#search_highlight0.

Radder, H. 1996. *In and About the World: Philosophical Studies of Science and Technology*. Albany: State University of New York Press.

Radder, H. 2006. *The World Observed/The World Conceived*. Pittsburgh, PA: University of Pittsburgh Press.

Radder, H. 2016. *Er middenin! Hoe filosofie maatschappelijk relevant kan zijn*. Amsterdam: Uitgeverij Vesuvius.

Radder, H. 2019. *From Commodification to the Common Good: Reconstructing Science, Technology, and Society*. Pittsburgh, PA: University of Pittsburgh Press.

Reiss, J. 2019. "Expertise, Agreement, and the Nature of Social Scientific Facts or: Against Epistocracy." *Social Epistemology* 33 (2):183–192.

Smolka, M. 2019. "*Studying Emotions in Meditation Research: Ethics and Epistemology Entangled?*" Paper presented at the conference on Bias in AI and Neuroscience, Full Program, pp. 106–111. Radboud University, Nijmegen.

Stone, I.F. 1989. *The Trial of Socrates*. New York: Anchor Books.

Van Hintum, M. 2015. *Wat wil Nederland weten? De totstandkoming van de Nationale Wetenschapsagenda*. Amsterdam: Nijgh & Van Ditmar.

7 Participatory Democracy and Multi-strategic Research

Hugh Lacey

SWARTHMORE COLLEGE/UNIVERSITY OF SÃO PAULO

List of acronyms used frequently in the text:

CSs	context-sensitive strategies
DSs	decontextualizing strategies
DS-research	research conducted under DSs
MS-research	multi-strategic research (research that may be conducted under CSs as well as DSs)
PD	participatory democracy
RD	representative democracy
$V_{C\&M}$	values of capital and the market
V_{TP}	values of technological progress

7.1 Introduction

What approaches to science should be fostered in the light of holding democratic values? Answers to this question differ depending on how "democracy" and "science" are interpreted.

I will sketch two interpretations of democracy: *representative democracy* and *participatory democracy*; and two (not so familiar ones) of what constitutes scientific research: *decontextualizing research* (DS-research), i.e., systematic empirical inquiry conducted under *decontextualizing strategies* (DSs), and *multi-strategic research* (MS-research). DS-research enables entertaining and confirming hypotheses and theories that have to do with the underlying structures of phenomena/objects/systems, of their constituents and their processes and interactions, the causal networks they belong to, and the laws governing them, considered in dissociation from their human, social, and environmental contexts, and that are tested in view of their fit with relevant empirical data that are normally obtained in the course of measurement (instrumental) and/or experimental interventions. In MS-research, DSs are complemented by *context-sensitive strategies* (CSs). Empirical investigation of phenomena that are inseparable from their human, social, and environmental contexts (see Section 7.2.4) require adopting some CSs and, under them, claims are tested in view of their fit with empirical data that may be qualitative and/or involve interpretations of natural and human phenomena in their ecological and social contexts. DS- and MS-research reflect often competing views of admissible (or privileged) methodologies and the priorities of scientific research, and they are associated with disagreements about what items of scientific knowledge may legitimately be used to inform social practices and daily life.[1]

My principle objective is to show (in Section 7.3) that holding the values of participatory democracy leads to fostering MS-research. First, however, I will discuss connections between representative democracy and DS-research, and the consequences that may follow for science from threats being made in several countries today to weaken democratic values and institutions.

7.2 Representative Democracy and Decontextualizing Research

7.2.1 Representative Democracy

Representative democracies (RDs) – the variety of political systems, self-identified as democratic, that function in the USA and many other nations – are marked, if only ideally or according to articles of their constitutions, by the separation of powers, networks of governmental bodies with responsibilities ranging from the local, through the provincial to the national, representatives in legislative and (in many cases) executive bodies at most levels chosen in elections in which all eligible citizens can vote, freedom of the press, the rule of law, and juridical protection of civil/political, property, and (in some cases) the full array of economic/social/cultural rights.

Democratic values are identified in RDs as the values that are articulated (in constitutions, by political parties, and legal and scholarly commentators) as shaping their core institutions, even if actually it is only a remote ideal that the values be robustly embodied in them. They include individual freedom, equality of opportunity, equality before the law, and protections for wealth and property, which in most RDs are very highly ranked values, even in those whose constitutions articulate respect for equality and solidarity. How these values are embodied varies among the RDs, and their embodiment can be enhanced or, as is happening today in the USA, Brazil, and several other countries, that of some of them can be diminished. Their robust embodiment is generally held to fit with the ideal that all people have the opportunity to live in ways experienced as being reasonably fulfilling, and so have access to the material conditions (including employment) to do so. There are disagreements, however, about whether to provide welfare programs for those who lack (or do not avail themselves of the opportunity to gain) access to these conditions. Entrepreneurial initiatives tend to be valued in RDs. And, in most of them, so are (with varying degrees of government tolerance and support) a vast array of independent civil society organizations and social movements, representing a pluralism of value outlooks, that deal with matters – connected with education, health, protection of human rights, recreation, art, sport, care for children, the disabled, and the elderly, charities, special interest groups, etc. – that may fall outside a government's purview or conflict with its policies. Valuing some of these organizations and movements tends to be reinforced by the conviction that adherence to democratic values is nourished by participating in them.

It is also widely believed in most actual RDs that commitment to the values of technological progress (V_{TP}) contributes to (and safeguards) the robust embodiment of democratic values, where commitment to V_{TP} involves (in summary) according high ethical and social value to exercising technological control over natural objects, to expanding human capacities to exercise such control in more and more domains of human and social life, and to the definition of human, social, and ecological problems in terms that permit solutions that deploy innovations derived from DS-research (Lacey 2005, 18–24). In addition, since institutions and interests that embody values of capital and the market ($V_{C\&M}$) (i.e., values such as economic growth, private control of productive forces, competition, profit, and private property) are the principal bearers of V_{TP} today, the prospects for strengthening democratic values are widely believed to be linked with furthering the trajectory being shaped by V_{TP} and $V_{C\&M}$. Then, V_{TP} and $V_{C\&M}$ tend to be considered inseparable from the core democratic values and integral to how they are interpreted.

Thus, the pluralism of value outlooks fostered in contemporary RDs tends to be bounded by the possibility of achieving an acceptable balance among

1   strengthening the core democratic values and institutions,
2   encouraging and tolerating a range of independent civil society organizations, and
3   commitment to V_{TP} and $V_{C\&M}$.

Political parties compete about how to interpret these items, rank them in order of importance, and determine what counts as an acceptable balance among them. In many RDs there is a virtual consensus among the leading political parties that an acceptable balance depends on interpreting (3) to require that decisions made in certain areas (often justified by appeal to the judgments of economic and scientific experts) should prioritize, or at least not be opposed to, relevant interests of property, capital, and the market. These areas include the production and distribution of good and services; the goals and processes of the workplace and farming practices; the use of natural resources and goods; the kinds of social arrangements that may exist and flourish; and how to balance institutionally social/economic/cultural rights with civil/political rights.

7.2.2 *Science in Representative Democracies*

For many decades following World War II, many policy makers in RDs accepted that not only does applying scientific knowledge contribute to building strong military capabilities, to enhancing the lives of most people, and to strengthening the institutions on which a vibrant RD depends, but also the development of science itself represents one of the great achievements of human history that any progressive society should foster. Hence, for them, it is right and normal for RDs to recognize the authority of well-confirmed scientific results, grant a significant degree of autonomy to science, require science to be a part of general education, and support with public funds research conducted in independent scientific institutions.[2] They generally thought of science as that which is practiced by qualified professionals in institutions publicly certified as "scientific", taught in university and school science programs, and reported in the science pages of the mainstream press, and its values to be those articulated, if not always well embodied, in these practices and institutions. The research conducted in them utilizes to some extent a variety of empirically based methodologies as appropriate in view of the characteristics of the objects/phenomena being investigated (and those of the human sciences are not usually considered reducible to those of the natural sciences). When explicitly articulated, however, scientific research tends to be thought of in terms of idealized features of the research practices that led to the great discoveries of the natural sciences, notably those that for the most part utilize methodologies that involve the adoption of DSs (Lacey 1999; 2019c).

RDs have normally provided substantial support for DS-research in ways that allow a significant measure of autonomy[3] to scientific institutions, and policy making in them is typically informed by the authoritative outcomes of this research. DS-research, by expanding our understanding of the underlying mechanisms and laws of phenomena, certainly generates the kind of understanding that leads to technoscientific discoveries and highly valued innovations in fields such as communications, medicine, agriculture, armaments, energy, and transport, many of which enhance the quality of our lives, widen the horizon of our aspirations, and open new possibilities for progressively reorganizing societies. These innovations and the practices and institutions in which they are used nowadays normally embody V_{TP} and $V_{C\&M}$, at the same time that they embody values connected with health, communications, etc., so that even when DS-research is conducted for the sake of (e.g.) dealing with a health problem, it also tends to enhance the social embodiment of V_{TP} and $V_{C\&M}$. Reflecting this, a fourth item, (4): fostering the pursuit of DS-research, may be appended to the items (1) – (3), among which, when appropriately ranked in order of importance, an appropriate balance is sought in RDs:

1   strengthening the core democratic values and institutions,
2   encouraging and tolerating a range of independent civil society organizations,
3   commitment to V_{TP} and $V_{C\&M}$, and
4   fostering the pursuit of DS-research.

That does not mean that DS-research is always conducted *for the sake of* generating technological innovations. Many of the discoveries of modern science have been made in the course of "basic research", i.e., research conducted with the aim to expand knowledge of the underlying order of phenomena, and not directly to generate efficacious applications that strengthen V_{TP} and $V_{C\&M}$. In addition, DS-research may be conducted for the sake of innovating in areas like medicine, where it may be considered just a contingency of the current situation of research and development that most innovations are likely also to embody $V_{C\&M}$. And there are DS-research projects, like those that provide the empirical and theoretical bases of the IPCC (Intergovernmental Panel on Climate Change) reports, that are conducted *for the sake of* obtaining understanding of phenomena – e.g., the cumulative socially and environmentally catastrophic effects of using technological innovations that depend on burning fossil fuels – that may point to the need to impede (or redirect) the trajectory being shaped by interests that embody V_{TP} and $V_{C\&M}$. This creates tension in contemporary RDs and scientific institutions, and deepens conflicts related to the order of importance of the items (1)–(4), and the appropriate balance to aim for among them.

There are additional sources of tension and conflict. Scientific institutions (and especially their funding sources) today increasingly tend to prioritise *technoscience*, i.e., DS-research that *aims more or less directly* to generate technoscientific innovations (Lacey 2012; 2016; 2020), and (many of them) *commercially oriented technoscience*, i.e., technoscience in which the innovations developed for use in the areas mentioned above are *intended* to serve interests that embody $V_{C\&M}$. [4] When these tendencies are accompanied (and fostered) by the election of authoritarian-leaning politicians and parties to head governments, as recently has happened in several RDs including the USA and Brazil, challenges emerge to the authority and autonomy of science, and sound scientific input into the formation of public policies (especially in matters connected with climate change and its causes and with the risks of using many technoscientific innovations) is often rejected; and efforts are made to subordinate all the items (1)–(4) to economic growth and other values of $V_{C\&M}$ (and perhaps also to personal interests of government officials and their political and electoral base).

7.2.3 Weakened Democratic Institutions and Commercially Oriented Technoscience

The governments in question exercise power in ways that weaken democratic values and institutions, and, in alliance with their corporate allies, provide backing for efforts to *identify* science with commercially oriented technoscience and to reject the authority of science (even DS-research) insofar as it extends beyond it. They interfere intrusively into scientific activities, aiming to subordinate the conduct and results of certain areas of scientific research to their interests (or whims) and/or those of their allies that embody $V_{C\&M}$, and to exclude undesired research and its results from the science curriculum in schools and universities as well as from having a role in informing public policies. These areas include investigations of the full range of human/social/cultural/ecological risks of (and harm already caused by) using agrotoxics in agricultural practices, as well as of many mining, drilling, hydroelectric, road-building, and other "development" projects in fragile environments, and certain modes of storage of waste products of mining, electricity generation, and industrial activities.[5]

The most prominent area has to do with investigations of global warming and climate change. The authoritarian-leaning governments and many of their allies deny the key conclusion of IPCC-related research (notwithstanding that *for the most part* it derives from DS-research): that global warming and currently observed and predicted climate changes are caused principally by the widespread use of greenhouse-gas-emitting technological objects and practices (and making available the conditions for their use) as well as the demise of the spaces (e.g., flourishing tropical forests) in which these gases can be readily absorbed. Moreover, the denial is accompanied by rejecting IPCC's and related proposals (that may be informed in part by knowledge obtained in MS-research conducted under CSs) for mitigating and adapting to the harmful effects of global warming and climate change on human beings, society, cultures, and ecological systems. The denial has practically nothing to do with critical appraisal of the data and models presented in the IPCC reports.[6] Moreover, the means utilized to make it appear plausible are incompatible with respecting the ideals articulated by the modern scientific tradition (see Section 7.2.4), as well as with adhering to the core democratic values. They include attacks on scientists who conduct the research: malicious insinuations about their scientific credentials, competence, and integrity; portrayals of them as purveying "false facts" for ideological reasons; intimidation and threats of taking legal action against them; dismissing them from employment (especially in government agencies) or from roles in public commissions; censoring publications and professional presentations; and other "punishments", e.g., salary reduction, denial of entitled promotions, and withholding funds for conducting research.

The interference is instigated principally by powerful and influential corporations in the energy, transport, logging, mining, agribusiness, chemical, and some other sectors (with the backing of politicians, governments, and scientists that support them). These corporations have a big share of the responsibility for the build-up of greenhouse gases in the atmosphere (as well as for risks in other areas occasioned by some uses of technoscientific objects) that are already leading to severely destructive consequences, such as more intense and perhaps more frequent hurricanes, floods, droughts, and bushfires, rising sea levels, and large-scale migrations from impoverished, climate-devastated regions. Scientific research, which investigates these harmful effects and risks of additional ones occurring, is targeted for interference, for its results cast into doubt the ethical/social legitimacy of actual uses of these objects and maintaining the conditions and policies that enable them. The interference is aimed at preventing knowledge about the risks of relevant corporate activities from being taken seriously (or even obtained) and diverting attention from the causes of the harm that has already occurred, as well as undermining any impact that knowledge might have on public policy, whenever possible making sure that financial and institutional support is not available for conducting the relevant research so that the grounds for raising the questions about legitimacy are unable to obtain strong empirical confirmation. Suppressing knowledge on these matters, and undermining research efforts to obtain more of it, are also important for efforts being made by the recently elected authoritarian-leaning governments to reduce opposition to such objectives as, in Brazil, privatizing universities and research institutions and clearing of much of the Amazon forest in the name of economic growth and increasing exports; and, in the USA, withdrawal from the Paris Agreement on matters relating to greenhouse gas emissions, weakening regulations that affect fossil-fuel-consuming industries, and providing large subsidies for agribusiness.

Interference *of the kind just described* is directed toward scientific research (including DS-research) insofar as it goes beyond commercially oriented technoscience; but the latter's scientific authority is unchallenged and its role in public policy formation considered indispensable. Technoscientific innovations provide the cutting edge of the trajectory of powerful interests (of corporations in the energy, transport, chemical, agribusiness, etc. sectors, not only of authoritarian-leaning governments). These interests aim to subordinate all values and the items (1)–(4) to $V_{C\&M}$, and so they effectively identify scientific research with commercially oriented technoscience, which unfolds today largely shaped by them without the interference described above, and often supported by public policies that provide privileged funding for research that is expected to lead to outcomes that serve them well.

The governments in question generally aim to privatize as much as possible, and so they may reduce public financing for scientific research, while at the same time providing incentives for the interests – connected with energy, transport, agribusiness, etc. that currently play the dominant role in shaping the trajectory that furthers $V_{C\&M}$ – to compensate by themselves becoming more involved in financing research in commercially oriented technoscience and providing employment possibilities for researchers in the laboratories they operate or sponsor. Hence, this research continues without interruption, not autonomously but under the constraints provided by the private funders and directed to the specific areas of research they choose.[7] Thus, the interference and the accompanying support for commercially oriented technoscience favor powerful special interests and defend them against competitors, e.g., proponents and developers of "green technologies" and supporters of the "green new deal", whose products and practices do not incur the same kinds of risks and that may challenge the powerful role that the special interests have come to assume in contemporary RDs. Sometimes they are portrayed, however, not so much as defending these special interests, but as defending the necessary role that commitment to $V_{C\&M}$ plays in RDs against threats derived from alleged misuses of the authority of science that are occasioned when public policies are informed by the results of the IPCC reports and related research.

This portrayal cannot pass scrutiny. Prioritizing the development and implementation of green technologies does not per se threaten commitment to $V_{C\&M}$. Its proponents do aim to develop technoscientific innovations that can contribute to mitigating or adapting to the effects of global warming, and to reshaping economic and other institutions so that they are not dependent on technologies that involve emitting greenhouse gases. However, many of them regard this as modifying, not discarding, the trajectory of V_{TP}, and contributing to strengthening the trajectory of $V_{C\&M}$ whose long-term viability, they maintain, depends on cutting down on the use of greenhouse-gas-generating technologies. Research aiming to generate green technology innovations is mainly conducted under DSs; and its proponents tend to think of scientific research principally in terms of DS-research (but not limited to commercially oriented technoscience), and maintain that the authority of DS-research should be recognized within RDs and policies enacted that respond appropriately to its input. However, since they aim to further V_{TP} and $V_{C\&M}$ in modern societies in ways that underlie a better balance of sustainability with the items (1)–(4),[8] they do threaten the dominant place occupied in RDs by special (oil, coal, chemical, agribusiness, etc.) interests that tend to interpret item (3), commitment to V_{TP} and $V_{C\&M}$, in terms that involve supporting the trajectory of V_{TP} and $V_{C\&M}$ shaped by their own interests.

Other competitors, including many who endorse the "green new deal", question whether the problems of climate change can be adequately redressed by focusing primarily on the quest for technoscientific innovations that reduce greenhouse gas emissions and global warming, and changes in the economy, transport systems, etc. to accommodate them. They often claim that, if the changes are to be consistent with upholding social justice and protecting the rights of everyone, more fundamental changes of lifestyle and values are also needed and that the relevant technoscientific innovations need to be adapted to them (Klein 2019). This claim can be investigated only when relevant CSs are adopted. Endorsing it would entail not only rejecting the dominant influence in RDs of the oil, coal, automobile, chemical, agribusiness, etc. interests, but also fostering the conduct of MS-research (in which CSs as well as DSs are adopted) on socially significant questions,[9] and subordinating V_{TP} and $V_{C\&M}$ to the values of social justice and protecting the rights of everyone and other values of participatory democracy (see Section 7.3.1). In Section 7.2.2, I introduced a list of items among which, when appropriately ranked in order of importance, an appropriate balance is sought in RDs:

1   strengthening the core democratic values and institutions,
2   encouraging and tolerating a range of independent civil society organizations,
3   commitment to V_{TP} and $V_{C\&M}$, and
4   fostering the pursuit of DS-research.

Endorsing the claim implies changing (3) to (3a), and (4) to (4a):

3a subordinating commitment to V_{TP} and $V_{C\&M}$ to upholding the values of participatory democracy,
4a fostering the conduct of MS-research on socially significant questions.

As maintained above, interfering with science is taking place side by side with attacks on democratic values and institutions in some RDs by those who interpret the aspiration to achieve an appropriate balance among (1)–(4) to require subordinating all the items to the furtherance of $V_{C\&M}$. These attacks lead to reducing RD to formal election processes, weakening measures that protect the rights of citizens, including the right to vote freely without being subject to manipulation and to participate in civil society groups and movements; and the attacks on scientists are themselves indicative of threats to their rights, e.g., to free expression of ideas and workplace security. When RDs are weakened in this way and authoritarian-leaning governments are elected, item (4) effectively becomes modified to

4b fostering the pursuit of commercially oriented technoscience, and rejecting the authority of DS-research insofar as in goes beyond it.

Both the embodiment of democratic values and the scope, autonomy, and authority of DS-research are thereby further weakened.

7.2.4 The Importance of Multi-strategic Research

Claims about what kind of balance *can* be achieved among the items (1), (2), (3/3a), and (4/4a/4b), and how it may be affected by subordinating V_{TP} and other values to $V_{C\&M}$ (or not doing so), including the claim that subordinating all the items to $V_{C\&M}$ is necessary for the future viability of RDs, are open to empirical investigation, the outcomes of which should inform what the various protagonists consider to be a *desirable* balance among the items. Investigating them involves (among other things) entertaining hypotheses, e.g., about the social embodiment of values, and so needs to deploy categories that are unavailable under DSs but available under some CSs. Thus, when science is limited to commercially oriented technoscience, it lacks the conceptual resources to empirically investigate the claim that subordinating all the items to $V_{C\&M}$ is necessary for the future viability of RDs. Thus, when this claim is endorsed by those who limit science in this way, it cannot be on the grounds that it is well confirmed in scientific investigation. Rather, it functions as a dogmatic presupposition of the legitimacy of the current trajectory shaped by $V_{C\&M}$, whose defense depends on the kinds of intrusion into scientific activities discussed above.

Spokespersons of RDs normally affirm the value of policy making being informed to the extent possible by relevant well-confirmed scientific results that generally are outcomes of DS-research, even if that value is not always manifested in practice (see Section 7.2.2). However, by not engaging in MS-research, in which relevant CSs are adopted, the claims referred to in the previous paragraph cannot be adequately investigated; and, when they are not, policies with profound social and environmental implications are introduced without being informed by adequately investigated claims. This points to a reason to change (4) to (4a), as well as to explore the potential consequences of replacing (3) by (3a).

Engaging in MS-research is further motivated in light of the fact that CSs (as well as DSs) must be adopted in investigations, not only about the claims just discussed, but always when investigating phenomena that cannot be separated from their environmental, human, and social *contexts*. This includes research on the full range of risks and harm actually occasioned by using technoscientific innovations,[10] the benefits of using specific technoscientific objects and the fairness of their distribution, and alternative practices (e.g., agroecology) that aim to integrate productivity with environmental sustainability and human/social well-being and subordinate introducing and using technoscientific innovations to this aim.[11] The need to adopt CSs in these investigations makes clear that DS-research by itself cannot be robustly responsive to the ideals transmitted throughout the modern scientific tradition that I have called *impartiality, inclusivity and evenhandedness, comprehensiveness*, and *autonomy*.[12] It cannot be responsive to *comprehensiveness*, e.g., since it cannot adequately deal with all the relevant phenomena connected with risks, benefits, and alternatives; or to *inclusivity and evenhandedness*, since there are value outlooks (e.g., that of participatory democracy) for which these phenomena are highly ethically and socially salient (Lacey 2019a).[13] Hence, to be responsive to these ideals requires going beyond DS-research and engaging in MS-research.

MS-research is especially significant for those who perceive the world's trajectory – fueled by technoscientific innovation and shaped by the currently dominant interests that highly embody $V_{C\&M}$ and subordinate the core democratic values (and science) to them – to be heading toward disaster. The disaster is not only connected with climate change and threats of social and environmental devastation, but also with the undermining of the conditions needed for fulfilling lives for vast numbers of people, for maintaining communities of traditional and working peoples, and for remediating the conditions especially of those who suffer from the residue (and recent intensifications) of racist policies and institutions. It is creating today's "crisis of democracy". This crisis has led some people to despair (and its accompaniments like drug addiction), nostalgia for a lost past, anger toward those perceived as causing this loss, migration, or openness to the appeal of authoritarian politicians. It is drawing others into movements of participatory democracy.

7.3 Participatory Democracy and Multi-strategic Research

7.3.1 Participatory Democracy

Participatory democracy (PD)[14] is rooted in the aspiration that all human beings exercise their agency as fully as possible – that conditions become available so that people and their communities are enabled to cultivate and express the values that they authentically embrace as integral to lives that express human well-being, as well as to act in the light of their own judgments intelligently informed by knowledge obtained in relevant inquiries, especially concerning matters that have significant impact on their lives and the possibilities they afford.

Authentically embraced values, as distinct from values accepted just for the sake of adapting within the prevailing social institutions and structures, derive from experience, tradition, reflection, and critical dialogue that draw upon the conceptions of nature, human nature, and proper relations between human beings and nature that make sense of people's conceptions of well-being (Lacey and Schwartz, 1996). And pertinent matters having significant impact on their lives include those found in the areas listed at the end of Section 7.2.1: the production and distribution of good and services; the goals and processes of the workplace and farming practices; the use of natural resources and goods; the kinds of social arrangements that may exist and flourish; and how to balance institutionally social/economic/cultural rights with civil/political rights.

The aspiration that *all* human beings exercise their agency as fully as possible underlies the values upheld by PD, values like the following (Lacey 1999, chapter 8; 2015a): solidarity in balance with respect for individual and group differences; the well-being and flourishing of everyone everywhere – creating a society in which all people, groups, and communities can live well and freely in accordance with their own authentically embraced values, and in which the enhanced agency of everyone is cultivated; giving high priority to the rights of the marginalized and exploited, and paying attention to the rights of the young and future generations in the context of a balance of civil/political with social/economic/cultural rights that is judged acceptable in the course of deliberations with widespread participation. These values, in turn, are dialectically linked with (among others) respect for nature, ecological/environmental sustainability, maintenance of biodiversity, and caution in the face of risks to health and environment. PD aims to subordinate V_{TP} and $V_{C\&M}$ – and the individualist values, whose embodiment cannot be universalized, generally fostered within RDs – to values like these.[15] What values are upheld by PD is always open to contestation and reconsideration in the light of ongoing reflection, critical dialogue, and changing circumstances.

Moreover, the values may be held with varying emphases, rankings, and culturally specific articulations; and they are compatible with a society that includes a pluralism of ways of living (but not a pluralism in which the exercise of agency of some undermines that of others).

PD is grounded in the groups (programs, movements, and organizations)[16] that embody these values now to a substantial degree in their practices and ways of living and that, by virtue of embodying them, represent in anticipation a type of society that would more robustly and widely express its aspirations. These groups are essential causal factors – "seeds" – of the struggle toward the more complete embodiment of the values in new political and socioeconomic arrangements (Lacey 2015a). That struggle involves "growth", typically in the "soil" provided by the constitutional provisions of RDs, whose institutions it aims to "fertilize" with the values of PD, e.g., by replacing items (3) and (4) respectively by (3a) and (4a) in the list of items among which, when appropriately ranked in order of importance, an appropriate balance is sought in RDs (see Section 7.2.3). Since it retains items (1) and (2), PD is not antithetical to the core values and institutions of RD.

How PD unfolds is not determined by governments (or powerful special interests), although they can contribute by creating spaces, conditions, and policies propitious for its more ample development, e.g., by giving subsidies to agroecology, rather than agribusiness and the farmers connected with it. The appeal of PD is grounded in the partial realizations actually achieved in the groups (movements and communities) that provide concrete images of alternative ways of living that incorporate the values, and its current expressions do not derive their value fundamentally from any role they might have as means for the realization of a theoretically characterized (albeit not yet existing) form of society. PD is nourished *now* for its adherents by the measure of well-being and effective agency that they experience within their communities and movements, and that they hope can draw ever more people to participate. The struggle of and for PD represents an "organic" or "dialectical" unity between means and ends, between ameliorative action and structural transformation, between short- and long-term objectives, and between personal development and social change. The measure of its degree of realization cannot be technological progress or economic growth; only the wider and deeper embodiment of its values can be.[17]

7.3.2 Links between Participatory Democracy and Multi-strategic Research

PD tends to be dismissed as, among other things, "utopian", "unrealistic", and "not recognizing the facts of human nature", and so to be open to cooptation by authoritarian forces that represent its antithesis. These charges, and assessing the possibilities for the "growth" of PD, cannot be adequately addressed without investigating the practices engaged in by the adherents of PD – the "seeds" – and appraising their possibilities, and to do this one must take into account how those practices can be informed by scientific knowledge. MS-research is indispensable here. The "seeds" are among the alternatives, which aim to integrate productivity with sustainability and human well-being and subordinate introducing technoscientific innovations to this aim (Section 7.2.4), and whose dynamic and possibilities cannot be adequately investigated in DS-research.

There are strong links between PD and MS-research: commitment to the values of PD fosters engagement in MS-research; and engagement in MS-research and the successful conduct of practices informed by knowledge obtained in it can strengthen interests that embody these values and so contribute to the growth of PD. It is beyond the scope of this chapter to discuss these links in general terms. Instead, I will focus on one of the "seeds" of PD, agroecology, which exemplifies how widespread popular participation in significant social practices can strengthen the agency of farmers and their communities.[18] Agroecology plays a central role in the social movements of the international network *La Via Campesina*, which propose the policies and practices of "food sovereignty" (Lacey 2015a; Wittman, Desmarais, and Wiebe 2010).[19] In recent years, various international bodies have endorsed the importance of agroecology and food sovereignty, both for addressing the food and nutritional needs of impoverished rural people in many parts of the world, and as an approach to agriculture that contributes to mitigating and adapting to the impact of agribusiness-promoted agriculture on global warming and climate change (de Shutter 2014).[20] I will briefly discuss the links between engaging in agroecology and adopting certain kinds of CSs in research.

7.3.3 Agroecology

Agroecology may be considered as an *approach to farming*, an *area of scientific research* (that deploys MS-research), a *social movement* (that embodies values of PD), and a *political project*.[21]

All four dimensions (components) of agroecology interpenetrate, reflecting *ways of interpreting the world* that may be represented in worldviews that vary considerably with the historical experience of peoples and across cultures in the styles, idioms, medias, metaphors, and ontologies utilized. These ways cultivate awareness of the many dimensions of things and practices, of the systemic and fluctuating connections among them, and the holistic unities (agroecosystems) of which they are components, as well as of the ecological and social conditions required for the different types of agricultural practices and interventions, the inseparability of environmental and social changes, the dialectical unity of natural processes and human ways of life, the coevolution of human cultures and nature, the importance of rejecting reductionist ontologies, and recognizing that (e.g.) seeds are objects of various types (not only biological, but also ecological and social objects) and that there are effects of using them occasioned by all of the types of objects that they are (Lacey 2020). These ways of interpreting the world also foster values of PD (like those listed in Section 7.3.1) and highlight human stances towards nature – respect, preserve, restore, sustain, cultivate, contemplate, appreciate, enjoy, love, harmonize with, mutually enhance, etc. – that (unlike the stance of unqualified control or domination that treats nature instrumentally and exploitatively) protect environmental sustainability, preserve biodiversity, and make it more likely that the regenerative powers of nature are not further undermined, and restored wherever possible. In addition, they promote a historical sense that resists the viewpoint that the trajectory of the hegemonic system (linked with agribusiness) is inevitable, and that looks to traditional/Indigenous practices as sources of ideas/ *saberes*/practices/collaborations of contemporary relevance.

Agroecology as approach to farming and area of scientific research will be discussed respectively in Sections 7.3.3.1 and 7.3.3.2.[22] As social movement, it is engaged in the struggle to consolidate, strengthen, and expand agroecology in all of its dimensions (or components). It opposes the political and economic arrangements present in most current RDs in which questions and public policies about the priorities of the production and distribution of agricultural goods, and the risks they occasion, are decided largely by agribusiness and the governments that support it on the advice of commissions of expert scientists who for the most part are engaged in commercially oriented technoscience. And it opposes that the deliberations of these commissions allow little room for the participation of, e.g., family, traditional, and Indigenous farmers, their organizations, and the input of knowledge generated under CSs that they bear, e.g., about the health problems, social disruption, and ecological devastation occasioned by agribusiness-sponsored farming practices that are dependent on chemical inputs and other technoscientific innovations. Its opposition on these matters derives largely from experience that shows that arrangements and deliberations of these kinds lead to undermining the social and environmental conditions needed for agroecological farming practices (and of PD), and often to disregarding the economic/social/cultural rights and weakening the civil/ political rights of family, traditional, and Indigenous farmers, as many of them are driven from the land into urban slums with precarious employment and educational possibilities and vulnerability to violence (Lacey 2017). It is not just an oppositional stance, however; it is combined with a positive political project that aims to implement policies (e.g., of land reform, social inclusion, and food sovereignty) that strengthen and expand networks of communities engaged in agroecology. It aspires in the long term to replace the current hegemonic food/ agricultural system with a new system (Lacey 2015a) that would be more sustainable (ecologically and socially), include an ensemble of farming practices urgently needed to adapt to or mitigate the effects of climate change (Perfecto, Vandermeer, and Wright 2019), and be more responsive to the food and nutritional needs of everyone in the world – and more generally to social justice, empowerment of the marginalized, and efforts to recuperate, strengthen, and develop the traditional knowledge and practices of rural and Indigenous peoples, as well as to efforts to organize constructive and mutually transforming interactions between these peoples and those whose lives and aspirations are highly informed by modern scientific knowledge (Lacey 2019b).

7.3.3.1 Agroecology as Approach to Farming

All farming takes place in *agroecosystems* that, broadly speaking, are geographically localized systems in which agricultural production takes place and its products are distributed. Agroecosystems may be thought of as historical, relatively enduring, transformable unities (systems or wholes) that may develop or degenerate, and that maintain a discernible identity that can be recognized over some period of time. They have social dimensions (production, use, and distribution of products), and are embedded in and not sharply separated from larger social systems, including networks of agroecosystems and the worldwide economic system, that affect their dynamics more or less intrusively and on which they have impact.

Agroecological farming is normally practiced by family, cooperative, and small-scale farmers; it utilizes organic and ecologically sustainable methods that use a minimum of external inputs, integrally takes into account the social and political dimensions of agroecosystems and embodies various of the values of PD, and is informed by "agroecological principles".[23] In contrast to the predominant practices of agribusiness-fostered agriculture that normally aim to *maximize* outcomes on a single or few dimensions (e.g., yields of a single crop or income), agroecological farming aims to bring about an *equilibrium* among outcomes on a variety of dimensions of the agroecosystems being cultivated. These dimensions may include: productivity, income, access to markets, variety and quality of products; robustness, resilience and adaptiveness, and conservation of biodiversity; strengthening of the agency, cultures, values, and self-reliance of farmers and of all members of their communities; social health, availability of healthy food, quality of life, and respect for the right to food security and other human rights; equal and just relations among men, women, and children; contributing toward bringing about and maintaining environments with non-polluted waters, fertile soils rich in microorganisms, and conditions available for preserving and expanding the use of locally relevant farmer-selected seeds, as well as social arrangements that foster such values as social inclusion, food sovereignty, the eradication of poverty, and the well-being of everyone everywhere. What are considered acceptably equilibrated agroecosystems and how that may change with changing social and environmental conditions, and which of their dimensions are to be prioritized in the equilibrium sought for, cannot be defined generally since it is determined principally, not by agricultural experts or supporting nongovernmental organizations, but by farmers in the agroecosystems being cultivated, their communities and organizations.

7.3.3.2 Agroecology as Area of Scientific Research

The general aim of agroecology as area of scientific research is to obtain understanding of agroecosystems, of their structures, constituents, functioning, dynamics of change, relations between human beings and natural objects in them, causal networks, and the impact of larger economic and climate-linked influences on them. This is the kind of understanding that can inform answers to the questions: what forms of intervention, practice, organization, and management enable the realization of the equilibrium that is considered acceptable by the farmers who cultivate an agroecosystem? And: under what conditions is an agroecosystem likely to degrade and what is the full range of human, social, and environmental consequences of its being degraded?

It is beyond the scope of this chapter to discuss agroecosystems in detail.[24] However, in view of their complexity, multi-dimensionality, variability from locale to locale, the places of human beings in them, and their holistic character, they cannot be dissociated from their human, social, and ecological contexts. Thus, in order to obtain understanding of them, an appropriate variety of CSs needs to be adopted in research. DS-research can lead to knowledge of objects/phenomena qua objects of the underlying causal order and qua objects of technical control, but not qua constituents of agroecosystems. Thus, it cannot deal adequately with the impact of using technoscientific objects on the realizability of the equilibrium desired of agroecosystems, or test the hypothesis that using them in the agroecosystems and under the socioeconomic conditions of their actual use might undermine the possibility of realizing it. For example, using seeds, qua objects of technical control (as when they have been modified by the techniques of genetic engineering) undermines the conditions needed to realize an acceptable equilibrium in agroecosystems that aim to strengthen the dimensions of sustainability, social health, and local agency (Lacey 2017; 2019c). Nevertheless, there are important roles for DS-research in agroecological research (and generally in MS-research): to investigate, e.g., constituents of agroecosystems such as minerals and microorganisms in soils; causal agents in them such as genetic structures and physiological and anatomical properties of plants; and diseases of plants and animals, and their causes. Moreover, using various technoscientific objects and procedures (but not GMOs) can contribute toward obtaining a desired equilibrium (Lacey 2015b; Nodari and Guerra 2015).

The required CSs may have some or all of the following features (Lacey 2015a): physical, chemical, biological, and ecological factors are intertwined with human, social, cultural, and historical ones; they incorporate what has been called the *diálogo de saberes*: knowledge and knowledge-gaining approaches of a multiplicity of mainstream scientific fields are integrated with those deployed by contemporary farmers in their activities and those developed by traditional and Indigenous farmers; claims and theories, proposed and tested under them, can deploy categories needed for understanding human agency, since human agents are among the essential constituents and causal agents of agroecosystems; and the knowledge obtained under them may play important roles in criticizing the hegemonic food-agricultural system.

Adopting these CSs and adhering to values of PD as embodied in agroecology, qua one of the "seeds" of PD, mutually reinforce each other. To show this it will suffice for present purposes to focus on the *diálogo de saberes* [25] and to point to three interconnected reasons for why it is essential for agroecological research.

First, in order to generate "local profiles". These chart the resources, dynamics of change, and possibilities of specific agroecosystems and, where they have been degraded as a consequence of the practices of agribusiness-fostered agriculture, the possibilities for transitioning to agroecology. They provide information about the knowledge and techniques used by farmers, the communities and movements they are part of, and the values they hold; catalogue objects that have economic, legal, cultural, aesthetic, cosmological, and religious value; and record memories and historical narratives that influence the definition of the equilibrium considered acceptable in the agroecosystems being cultivated. *Without access to local profiles agroecological principles cannot be applied in particular agroecosystems and the possibilities they afford assessed.* Sketching them requires taking into account the experience, knowledge, and practical activities of farmers, for they typically have a more complete grasp of the agroecosystems they cultivate than professional scientists could hope to obtain, including of the variety of their organic and inorganic constituents, of their historical patterns of variation, of the impact of climate change and of the harm that is occasioned when they become degraded, of the practices that can be sustained and maintain biodiversity in them, and of the languages, interests, capabilities, values, and aspirations of the people whose agency is to be exercised and strengthened by participating in the deliberations that lead to defining acceptably equilibrated agroecosystems. Moreover, their experience may well be informed by knowledge that has met "the test of time", e.g., about how to select seeds that are adapted to available agroecosystems, to conserve biodiversity for long periods of time under changing climatic conditions, and to recuperate degraded lands and forests (Lacey 2019b).

Second, the *diálogo* contributes to strengthen the agency of the farmers. It utilizes their knowledge, perceptiveness, and capacity to learn, to cooperate, and to make their own judgments and decisions. It is engaged in when conducting research – in which the farmers are key players in setting the objectives and evaluating how well they are met – that informs determining, developing, and maintaining desirably equilibrated agroecosystems. It encourages the recovery of ancestral farming knowledge (the loss of which is one of the casualties of the advance of technological and economic "progress", as well as of colonial conquest) and appropriating elements of agroecology that can strengthen their cultural and traditional heritages. It helps to shape lives with dignity for themselves and future generations of rural (including traditional and Indigenous) peoples, to provide a context for recognizing the contribution they are already making to providing a significant part of the food that is currently consumed, and to enable them to contribute positively to the solutions of the food, climate, political, and other crises that currently confront humanity.

Third, farmers are likely to have privileged access to much of the "empirical data" needed to test agroecological principles and knowledge claims. The needed data go beyond observational reports of such matters as the analysis of minerals in soils and the genomes and developmental processes of plants. They include reports of the observed intricacies of agroecosystems, of their impacts on human beings, of the aspirations, motivations, and hopes of farming communities and of how they define acceptably equilibrated agroecosystems. Professional scientists can access much of the data only through communicating with the farmers, whose observations and testimonies about these matters *ceteris paribus* carry epistemic authority. Thus, obtaining and evaluating claims made about agroecosystems and their possibilities require active and respectful communication with the farmers and, consequently, the rejection of asymmetric relations between scientific specialists and practicing farmers, the type of relations present when the objective of agricultural practices is considered that of maximizing results on a few dimensions.

7.4 Concluding Remarks

I have maintained that science, interpreted as MS-research, can contribute to strengthen PD, and that holding the values of PD can draw scientists to engage in MS-research and to criticize commercially oriented technoscience. Not only that: the ideals of the tradition of modern science – I have focused on *inclusivity and evenhandedness* and *comprehensiveness* – also can draw them into that engagement and criticism. Engaging in MS-research can contribute simultaneously to strengthen PD and to enable better accord with these ideals.

Although contemporary scientific institutions and research practices may not respond robustly to these ideals, their spokespersons often appeal to them. Furthermore, despite the growing tendency to identify science with commercially oriented technoscience, the science that is practiced in distinguished scientific institutions and taught in programs of science education and that has impact on policy deliberations continues (however precariously in some nations) to go well beyond commercially oriented technoscience. It includes not only a good amount of basic research, but also DS-research on the environmental impact of using certain kinds of technological developments (e.g., on global warming) that is of critical significance for the formation of policies that can maintain the viability of RD, and some research conducted under CSs in the ecological and human sciences.

Nevertheless, as outlined in Section 7.2.3, the tendencies to weaken RD and to identify science with commercially oriented technoscience have gained strength in several RDs, including in the USA and Brazil, and resulted in growing political interference into scientific research and education that is threatening the integrity of science. In order to defend democracy in any of its varieties, it is important to resist and denounce this kind of interference and to demand that ample public support be provided for scientific research that is not subordinated to the agendas of capital and the market and authoritarian-leaning politicians. That demand should be based not on the claim that scientific developments contribute indispensably to economic growth, but on the contribution that scientific research and its results can make (when responsive to all the ideals of the modern scientific tradition) to further the core democratic values.[26] Moreover, it should recognize the necessity of engaging in MS-research in order to understand better how the core democratic values can be connected with V_{TP} and $V_{C\&M}$ and with the values of PD.

The demand should be made in the name of the integrity of science and commitment to the idea that science belongs to the shared patrimony of humankind. While it is true that it may have little direct impact on authoritarian-leaning politicians, and others who give high salience to V_{TP} and $V_{C\&M}$, that is not a good reason to accept the submission of scientific research and education to their interests and requirements.[27] Moreover, maintaining the demand puts science at the side of (and strengthens) other forces in the weakened RDs, which are engaged in the struggle to uphold democratic values and recognize the role normally accorded to science in well-functioning RDs (see Section 7.2.2). Wherever the demand can be responded to, it is likely to contribute to instigating that space be opened for PD to unfold within the framework of RD. That could lead to conditions in which the authority of science would be strengthened, so that, when it is well exercised in MS-research, it would be able to influence the formation of public policies soundly and reliably, and thereby enhance the embodiment of the core values that are articulated in RDs and enriched in PD.

Notes

1 The distinction between DS-research and MS-research and the theses, that scientific research is always conducted under a strategy and that there are different kinds of strategies (DSs and CSs), are elaborated (sometimes using different terminology) in Lacey (1999; 2005; 2014; 2016; 2019a; 2019c); Lacey and Mariconda (2014).

2 This viewpoint (most famously put forward in Bush, 1945) was not unanimously accepted. In particular, a strong opposition persistently maintained that public support should only be provided for militarily or economically useful research projects.

3 The autonomy is bounded. Some government interference with science is widely considered legitimate, e.g., "classifying" certain research and its outcomes for the sake of defending national security or military secrets.

4 In Brazil and several other South American nations in recent years, government ministries of science and technology have been renamed ministries of science, technology, and innovation, and their websites make clear that their interest is in fostering commercially applicable innovations.

5 In these investigations, e.g., concerning the full range of risks of using agrotoxics and GMOs, there may be an important role for CSs. There are professional scientists who participate in the attacks on these investigations, their results, and the scientists who conduct them; they may draw on some version of the view that

scientific research should be identified with DS-research (Lacey 2017; 2019c). These scientists, of course, do not support efforts to exclude the teaching of evolution from the school curriculum that supporters of authoritarian-leaning politicians often support. Unlike the issues about risks, these efforts have nothing to do with strengthening $V_{C\&M}$. However, they pose no threat to the pursuit of commercially oriented technoscience, and thus no obstacles to political alliances being formed between their followers (religious fundamentalists) and adherents of neoliberal economic policies. Both groups have interest in impeding science insofar as it goes beyond commercially oriented technoscience.

6 Or with *evidence and arguments* that IPCC research is marred by bias, that the empirical data have been manipulated and altered when convenient, and that the "uncertainties" present in IPCC's predictions are so pervasive that a compelling case cannot be made now to act informed by them (see, e.g., Oreskes and Conway, 2010, chapter 6).

7 Patterns are different in different countries. The current Brazilian government has cut funds for all scientific research (and education) including in commercially oriented technoscience. It appears to be confident that, with the economy dominated more and more by international capital, it is more cost efficient for Brazil (and corporations located in Brazil) to count on access to research conducted in the USA and other countries and the innovations it generates. On the other hand, in (e.g.) the USA, many strong university-based research institutions maintain a considerable measure of autonomy although their research projects are increasingly being funded by private interests – but not without serious tensions emerging (see Michaels, 2020).

8 E.g., Carlos Nobre recommends that, in looking for new ways to redress the perils of global warming, we "aggressively research, develop, and scale a new high-tech innovation approach that sees the Amazon as a global public good of biological assets and biomimetic designs that can enable the creation of innovative high-value products, services, and platforms for current and for entirely new markets by applying a combination of advanced digital, material, and biological technology breakthroughs to their privileged biological and biomimetic assets" (Nobre et al., 2016).

9 Since MS-research includes research conducted under DSs, this would not imply downplaying the role of green technologies in addressing the issues connected with global warming.

10 An important question here concerns whether it is possible (and how) to introduce the technologies intended to mitigate global warming and to replace the greenhouse-gas-generating technologies – given the resources and conditions they require – without at the same time intensifying social injustice, e.g., by undermining the conditions needed for alternative practices and ways of life (e.g., Indigenous ones) that may not accord high salience to $V_{C\&M}$.

11 The CSs adopted in ecological investigation do not suffice for this research. They recognize a kind of complexity of ecological systems and the interaction of its components that can occasion the emergence of higher-level properties and permit two-way interactions among all levels of the system. Those needed for this research require, in addition, conceptual resources that can express that ecological systems are embedded in social systems, which are affected by events at all levels of the ecological systems, and that, in turn, social events may have causal impact at all of them. For the remainder of the chapter, when I refer to CSs, I will be referring to CSs with these properties.

12 See references in note 1. Leaving aside important nuances, *impartiality* is the ideal that judgments about the confirmation of scientific knowledge be made in the light only of relevant empirical data and evaluative criteria that are free from ethical/social value commitments. What I call *inclusivity and evenhandedness* is

one interpretation of what is often referred to as *neutrality*. It is intended to express the ideal that scientific research and knowledge belong to the shared patrimony of humanity, and so (in principle) can be of interest, more or less evenhandedly, to all value perspectives. *Comprehensiveness* is the ideal that all phenomena can in principle be grasped in scientific inquiry. A variety of complexities and difficulties arise in attempting to articulate the ideal of *autonomy* (see Lacey 1999, 248–254; 2016; 2019c).

13 Limiting scientific research to that conducted under DSs also leads to tensions with *impartiality*. For example, endorsing the claim that subordinating (1)–(4) to $V_{C\&M}$ is necessary for the future viability of RDs, on the grounds that it is a presupposition of the legitimacy of the current trajectory shaped by $V_{C\&M}$, is not in accord with *impartiality*. And *autonomy* is compromised when scientific institutions dismiss or marginalize the sound scientific credentials of MS-research (Lacey 2016; 2019c).

14 I have no space in this chapter to defend my characterization of PD or discuss how it fits with other accounts in the vast literature on the topic.

15 Of course, if a significant body of people does not authentically embrace these values (following reflection and critical dialogue), PD would not express a viable aspiration. Furthermore, as a referee pointed out, the viability of PD depends on the conceptions of nature and human nature, which underlie the values it aspires to embody more firmly, being in harmony with the way nature and human nature really are.

16 These may be among or in continuity with various of the independent civil society groups mentioned in the characterization of RD in Section 7.2.1, and they include many Indigenous and traditional groups.

17 Some authoritarian-leaning governments express aspirations whose degree of realization also cannot be measured by technological progress or economic growth but only by the wider and deepening of the embodiment of such values as the happiness and general well-being of their citizens. Even where these aspirations amount to more than verbal expressions, their realization is intended to be an outcome of top-down policies, not of the organic or dialectical unities that are indispensable to PD.

18 Other "seeds" of PD in Brazil include *"tecnologia social"*, a movement to foster the development and use of technologies that can be utilized in practices that aim for greater social inclusion and foster solidarity; movements connected with sustaining and maintaining the Amazon forest and defending the rights of Indigenous (and other traditional) peoples and affirming their agency; some feminist organizations; movements to open universities to previously excluded groups and to create university programs that teach and engage in research connected with Indigenous *saberes*; and groups aiming to develop community participation in public health services and alternative sources of energy that can serve the needs of poor communities. (Like the groups that foster agroecology, those that represent these "seeds" are threatened by policies and acts of Brazil's current authoritarian-leaning government.)

19 On food sovereignty, see https://viacampesina.org/en/food-sovereignty/;www.mstbrazil.org/news/statement-people%E2%80%99s-movement-assembly-food-sovereignty (accessed on December 14, 2019).

20 See also, e.g., Alliance for Food Sovereignty in Africa, https://afsafrica.org/case-studies-agroecology/ (accessed on April 7, 2020).

21 This summary statement draws upon statements of the groups in Brazil that constitute ANA (Articulação Nacional da Agroecologia, https://agroecologia.org.br/, accessed on April 7, 2020) and those articulated in the policies and practices of popular rural organizations (e.g., La Via Campesina, https://viacampesina.org/en/, accessed on April 7, 2020) in many parts of the world, and endorsed by

major theorists of agroecology and commentators on its history, especially Altieri (1995); Nodari and Guerra (2015); Perfecto, Vandermeer, and Wright (2019); Rosset and Altieri (2017). (The term "agroecology" is sometimes used to refer to agricultural approaches that do not incorporate all of the four stated dimensions. I do not discuss them here.)

22 For a more complete and less selective discussion, in which the interactions among all of the dimensions of agroecology are made more explicit, see the references in the previous note and Lacey (2015b).

23 Rosset and Altieri (2017, 20) list six of these principles. They include "[e]nhance the recycling of biomass, with a view to optimizing organic matter decomposition and nutrient cycling over time" and "[p]rovide the most favorable soil conditions for plant growth, particularly by managing organic matter and by enhancing soil biological activity". The principles are not outcomes of research conducted under DSs. Nevertheless, they are empirically well grounded in the experience of many farmers over the course of many years, and require for their application interpretation of, e.g., "optimizing organic matter decomposition" in contexts with very different soil components and conditions.

24 For more detailed discussion, including of the various kinds of constituents of agroecosystems, see Lacey (2015b), and the references in note 21.

25 The other features are discussed in Lacey (2015b; 2019b); and the notion of "*saber*", and the relationship between traditional *saberes* and scientific investigations and their results, in Lacey (2019b).

26 A referee maintained that, ironically, unless that is done, interests that embody $V_{C\&M}$ cannot be sustained in the long run, for these interests, if unrestrained and backed by the identification of science with commercially oriented science, "will extract, exploit, and deplete both natural and human resources and simply not be sustainable".

27 Of course, not accepting it can come with great personal costs and lead to further attacks on a nation's scientific institutions, so that demands made in the name of the integrity of science should be made with prudence, care, and courage, in a context in which incommensurable values may be in competition.

References

Altieri, M.A. 1995. *Agroecology: The Science of Sustainable Agriculture*. Boulder, CO: Westview.

Bush, V. 1945. *Science: The Endless Frontier*. Washington, DC: National Science Foundation.

de Schutter, O. 2014. *The Transformative Potential of the Right to Food*. Report of the Special Rapporteur on the Right to Food. Human Rights Council, General Assembly of the United Nations, January 24, 2014. Available online at www.srfood.org/images/stories/pdf/officialreports/20140310_finalreport_en.pdf. Accessed December 10, 2019.

Klein, N. 2019. *On Fire: The (Burning) Case for a New Green Deal*. New York: Simon & Schuster.

Lacey, H. 1999. *Is Science Value Free? Values and Scientific Understanding*. London: Routledge.

Lacey, H. 2005. *Values and Objectivity in Science*. Lanham, MD: Lexington Books.

Lacey, H. 2012. "Reflections on Science and Technoscience." *Scientiae Studia* 10: 103–128.

Lacey, H. 2014. "Tecnociência comercialmente orientada ou pesquisa multi-estratégica?" *Scientiae Studia* 14 (4): 669–695.

Lacey, H. 2015a. "Food and Agricultural Systems for the Future: Science, Emancipation and Human Flourishing." *Journal of Critical Realism* 14 (3): 272–286.

Lacey, H. 2015b. "Agroécologie: la science et les valeurs de la justice sociale, de la démocratie et de la durabilité." *Ecologie et Politique* 51: 27–40.

Lacey, H. 2016. "Science, Respect for Nature, and Human Well-being: Democratic Values and the Responsibilities of Scientists Today." *Foundations of Science* 21 (1): 883–914.

Lacey, H. 2017. "The Safety of Using Genetically Engineered Organisms: Empirical Evidence and Value Judgments." *Public Affairs Quarterly* 31 (4): 259–279.

Lacey, H. 2019a. "Roles for Values in Scientific Activities." *Axiomathes* 28 (6): 603–618.

Lacey, H. 2019b. "Ciência, valores, conhecimento tradicional/indígena e diálogo de saberes." *Desenvolvimento e Meio-ambiente* 50: 93–115.

Lacey, H. 2019c. "A View of Scientific Methodology as Source of Ignorance in the Controversies about Genetically Engineered Crops." In *Science and the Production of Ignorance: When the Quest for Knowledge is Thwarted*, edited by J. Kourany and M. Carrier, pp. 245–270. Cambridge, MA: MIT Press.

Lacey, H. 2020. "The Many Kinds of Objects that Technoscientific Objects Are." *Revista de Filosofia da Unisinos* 21 (1): 14–23.

Lacey, H., and P.R. Mariconda. 2014. "O modelo das interações entre as atividades científicas e os valores." *Scientiae Studia* 14 (4): 643–668.

Lacey, H., and B. Schwartz. 1996. "The Formation and Transformation of Values." In *The Philosophy of Psychology*, edited by W. O'Donohue and R.F. Kitchener, pp. 319–338. London: Sage.

Michaels, D. 2020. *The Triumph of Doubt: Dark Money and the Science of Deception.* New York: Oxford University Press.

Nobre, C.A., et al. 2016. "Land-use and Climate Change Risks in the Amazon and the Need of a Novel Sustainable Development Paradigm." *Proceedings of the National Academy of Sciences* 113 (39): 10759–10768.

Nodari, R.O., and M.P. Guerra. 2015. "A agroecologia: suas estratégias de pesquisa e sua relação dialética com os valores da sustentabilidade, justiça social e bem-estar humano." *Estudos Avançados* 15: 183–207.

Oreskes, N., and E. Conway. 2010. *Merchants of Doubt.* New York: Bloomsbury.

Perfecto, I., J. Vandermeer, and A. Wright. 2019. *Nature's Matrix: Linking Agriculture, Biodiversity, Conservation and Food Sovereignty.* 2nd edition. London and New York: Routledge.

Rosset, P.M., and M.A. Altieri. 2017. *Agroecology: Science and Politics.* Black Point, Nova Scotia: Fernwood Publishing Company.

Wittman, H., A.A. Desmarais, and N. Wiebe (eds.). 2010. *Food Sovereignty: Reconnecting Food, Nature and Community.* Oakland, CA: Food First.

8 Public Opinion, Democratic Legitimacy, and Epistemic Compromise

Dustin Olson

LUTHER COLLEGE, UNIVERSITY OF REGINA

8.1 Introduction

Currently, the United States offers a paradigm of partisan political theater. Nowhere is this partisanship on display more than with the reality and causes of global warming. As recently as 2008, however, high-profile Republicans Newt Gingrich (former Speaker of the US House of Representatives) and John McCain (former US senator and 2008 US presidential candidate) and high-profile Democrats Nancy Pelosi (current Speaker of the US House of Representatives) and Barrack Obama (44th US president) publicly agreed on the reality and dangers of climate change due to anthropogenic global warming (AGW). Within two years, climate change was one of the most divisive issues between the two parties. The political fissures surrounding climate change were to no minor extent the result of changing public opinion. A well-documented propaganda campaign against the realities of climate change by a number of vested-interest groups toward a number of voting demographics, particularly those in the Midwest most negatively impacted by free trade and the 2008 recession, served as a catalyst for this change. As a result, numerous incumbents, both Republican and Democrat, lost their seats in the 2010 mid-term elections, where Republicans took control of the US House of Representatives and gained six seats in the US Senate. In many cases, that a candidate accepted the reality of AGW was the primary reason for their loss. Due to the politicization of climate change and the public trend toward climate denial, policy makers from both political parties subsequently downplayed or even made an about-face on this issue. This absence is especially striking as anthropogenic climate change and the dire impacts implied by it was established science at that point.

Global warming due to CO_2 emissions is not a recent hypothesis. Indeed, Eunice Newton Foote (1856) and John Tyndale (1861) predicted this possibility well over a century ago. Moreover, in 1896 Swedish chemist Svante Arrhenius calculated that a doubling of CO_2 in the atmosphere would raise the average global temperature 5–6 degrees Celsius, "[a] conclusion," Isabel Hilton notes, "that millions of dollars' worth of research over the ensuing century hardly changed at all" (2008, 1). Into the 1970s, climate-related scientists began converging on the hypothesis that global warming would escalate if nothing was done to curb these emissions. By 1988, the data garnered sufficient support to bring public attention to the concerns. James Hansen, director at NASA's Goddard Institute for Space Studies, explained to the US senate,

> global warming is now large enough that we can ascribe with a high degree of confidence a cause and effect relationship to the greenhouse effect ... the greenhouse effect is real, it is coming soon, and it will have major effects on all peoples.
>
> (Hansen 1988, 39; rearranged)

By 2010, more than 97% of active climate scientists accepted that global warming was occurring at an increased rate due in large part to human activity (Cook et al. 2013).

Mitigating the effects of climate change is a leading candidate for the most pressing moral challenge in human history. Its contribution to the increased frequency of extreme weather, rising sea-levels, and extinction events is accurately described as a crisis. The decade since the 2010 US mid-term elections has been the hottest on record, with five of the ten hottest years occurring since 2015 (Lindsey 2020). The by-products of this sustained heating are not good. Increasing numbers of heat waves have killed thousands globally.[1] Wildfires and droughts have increased in number and intensity, with economic costs in the trillions, and immeasurable environmental tolls. In 2010, prolonged drought and sustained heatwaves killed in the tens of thousands. Mass graves were required in India in 2015 because morgues were overrun from the numbers of heat deaths. In 2018, wildfires in California razed 1.8 million hectares of forest. Through 2019 and into 2020 prolonged drought and sustained record-breaking temperatures in Australia resulted in wildfires scorching 17 million hectares of forest, killing 24 humans and nearly 1 billion animals in the process (UN 2020). Moreover, current projections suggest that 2020 will overtake 2016 as the hottest year in human history, which is sure to exacerbate the already catastrophic heatwaves, fires, floods, and super storms; in turn, this will exacerbate ice melt, sea-level rise, and mass migration. If left unchecked, climate change will continue to vastly and negatively alter human civilization away from its current form. If these sobering statistics do not lend evidence to climate change being *the* moral issue of our time, I am unclear on what the alternative could be.[2]

My concern here, however, is not to defend the moral imperative for mitigating AGW – I assume that this is the case without further discussion. Rather, I am concerned with the moral importance of what one believes insofar as one's beliefs influence one's actions, specifically in contemporary liberal democracies. We thus use the issue of climate change at this time in US politics as a specific example of more general challenges within these parameters. In particular, I am engaging in a social moral epistemology. My concern is with those institutions and practices influencing public opinion in ways antithetical to morally right action and how we might improve the conditions under which we can formulate accurate judgments on morally significant issues like AGW.[3] Between 2008 and 2010, for example, confidence among US citizens in the reality of anthropogenic climate change dropped from 51% to 33%, suggesting that nearly 70% of the populace were unsure of AGW's reality; the percentage of US citizens confident that climate change was *not* occurring increased from 9% to 11% in that same period (Leiserowitz et al. 2018). If indeed climate change is one of the most significant contemporary moral issues requiring sustained pooled cooperation, these numbers are unacceptable. When considering how to best deal with the challenges posed by anthropogenic climate change, denying the phenomenon that poses the challenge is a non-starter.

In what follows, I contend that the necessary division of epistemic labor, or our social epistemic interdependence, coupled with individual partisan bias are being exploited, leaving an epistemically compromised electorate. That is, members of an electorate are systematically inhibited in their abilities to form rationally considered political judgments, as they are subject to intentionally propagated distorting influences. Building from this assessment, I then consider the democratic legitimacy of decisions made under these epistemic conditions, arguing that the legitimacy of a democratic decision is challenged by an epistemically compromised electorate. The problem is not democracy, but rather the distorting influences used to undermine democratic decision making. Following John Dewey's (2010) observation that "unless freedom of action has intelligence and informed conviction back of it, its manifestation is almost sure to result in confusion and disorder" (126), I thus argue there are social moral epistemic grounds for mitigating what I dub social epistemic exploitation: the willful dissemination of distorting influences so others form judgments that will serve one's self-interests.

From this conclusion, I propose that we revisit our conception of traditional liberal institutions whose role is to facilitate public reason in light of the epistemic needs of democratic citizens. A minimal expectation for liberal democratic institutions, such as public education, journalism, and political campaigning, is that they do not undermine democratic legitimacy by corrupting public reason. As our example highlights, however, these institutions are in fact being exploited in ways that are corrupting public reason, ironically contravening their primary function. Proposing a fully worked-out strategy for how we might mitigate willfully propagated distorting influences and their effect on democracy is beyond the scope of this chapter. I do, however, conclude by offering some general considerations for how we might better structure those institutions designed for facilitating public reason given the epistemically corrupting influences currently emerging from them.

8.2 Epistemic Vulnerability and Epistemic Exploitation

Theorists from Plato to Dewey recognize the importance of pooled cooperation, specifically through participation in civic life. They stress, however, that vital to the success of such cooperation is that one's convictions are both informed and intelligent, features without which there will be more chaos than cooperation. One component of this pooled cooperation, and an inevitable condition of being an epistemic agent in a contemporary liberal democracy, is the level to which we rely on others when forming our beliefs. Most of our judgments on most subjects rely on information acquired from others' testimony. This reliance increases to the degree that the information in question is understood by only a specialized few, or is not easily accessible, or is too voluminous for one to reasonably consider. Our current circumstance is that there is simply too much information for one person to gather and assess for oneself, and on many issues, such as climate change, there are technical details requiring specialized knowledge. Given this reality, there is thus a necessary division of epistemic labor, which we can refer to as our *epistemic interdependence*.

One of the challenges emerging from our epistemic interdependence is that it is vulnerable to systematic exploitation. The inevitability of our epistemic interdependence requires that we must employ various sources to inform our political judgments. This circumstance leaves open the potential for information sources to cease communicating in good faith. Communicating in bad faith is to communicate under the pretense that one is concerned with the accuracy of one's testimony and with the goal of convincing another to accept that testimony. When one employs bad-faith testimony for the purposes of using someone else to achieve one's own ends, one is participating in what I dub *social epistemic exploitation* (SEE):

SEE A's assertion that P epistemically exploits B when

1   A asserts that P to influence B to act as if P is true;
2   B acting on the assumption that P is true benefits A;
3   A is indifferent to P's truth-value;
4   A's assertion that P is done in bad faith.

Examples of SEE include media sources that sensationalize an issue without appropriate proportionality or deference to epistemic authorities, with the goal of increased ratings or website "clicks"; corporate advertising and propaganda, with the goal of maintaining high profit margins; political candidates shilling for donations and votes; and financially motivated scientific experts playing the role of contrarian on behalf of business interests.

In cases of SEE, the speaker is not interested in communicating accurate information, but rather is solely concerned with convincing an audience that the content or connotation of the testimony is accurate. This point relates to Harry Frankfurt's (2005) conception of bullshit (BS), where a speaker offers

BS when asserting that P, but is indifferent to P's truth-value. It is helpful to note, however, that unlike SEE, not all BS is asserted in bad faith. A comedian may embellish a story to make an audience laugh, with no concern for the accuracy of the story's details. One would be remiss to accuse the comedian of communicating in bad faith, even if the story is BS. There is no pretense of communicating in good faith in this circumstance. This distinction highlights the type of bad-faith, or exploitative, communication I am concerned with. In cases of SEE the BS is asserted as not BS. This represents a particularly insidious class of BS, where there is the willful dissemination of distorting influences, influences that inhibit one's ability to form accurate judgments (Rawls 1971, §9), with the pretense of communicating in good faith.

The willful propagation of distorting influences, such as propaganda, bullshit, or outright lying is a common and effective tactic used to exploit our epistemic interdependence.[4] The effectiveness of these tactics is substantially amplified when they prey upon various biases in the target audience. Consider, for example, the significance of group identification over what we believe. Social psychologist Lillian Mason (2018) highlights the ways in which the social group to which we most closely identify establishes in us an emotional connection with the most widely accepted judgments associated with that group.[5] Our judgments consistently align with the group with whom we most closely identify and connect on an emotional rather than a rational level. A fortiori, once established, partisan biases have a much stronger effect on how and what we judge than the epistemic reasons, or evidence, we have favoring that judgment. Coupling the inevitability of partisan bias with distorting influences that intentionally target those biases and we have a recipe for an epistemically compromised electorate and a paradigm case of SEE.

We are thus epistemically vulnerable in two inevitable ways. First, our epistemic interdependence reveals a social epistemic vulnerability; second, our own partisan biases reveal an ego-centric epistemic vulnerability. Taken in tandem, the distorting influences employed in SEE are extremely powerful. Returning to our example, we find such tactics utilized to undermine the scientific consensus surrounding climate change into the 2010 US mid-term elections. Spearheaded by businesses and lobbyists with stakes in fossil fuel, there were concerted efforts to undermine public confidence in climate science on at least two fronts. First, these interest groups sought to cast doubt on the legitimacy of the science itself. Second, they sought to turn the reality of climate change into a political, rather than scientific, issue. Each of these strategies represents the ways distorting influences are strategically employed to trigger partisan bias and one way that an electorate can become, or remain, epistemically compromised.

8.2.1 Distorting the Science

Casting doubt on the science behind undesirable science-based policies is a not uncommon tactic used to undercut support for those policies.[6] Tim

Phillips, president of the Koch-funded Americans for Prosperity, emphasizes this strategy in relation to climate policy: "If we win the science argument, it's game, set, and match" (Hockenbury and Upin 2012). What constitutes "winning the science" under these parameters? Consider the stated goals of the Global Climate Science Communications Action Plan, offered by a conglomerate of fossil fuel businesspersons and lobbyists in response to the 1997 Kyoto protocol.

Victory will be achieved when

- Average citizens "understand" (recognize) uncertainties in climate science; recognition of uncertainties becomes part of "conventional wisdom";
- Media "understands" (recognizes) uncertainties in climate science;
- Media coverage reflects balance on climate science and recognition of the validity of viewpoints that challenge the current "conventional wisdom"; ...
- Those promoting the Kyoto treaty on the basis of the extant science appear to be out of touch with reality.

(American Petroleum Institute 1998)

According to Myron Bell, one of the co-authors of this action plan, "if you concede that the science is settled and there is consensus ... the moral high ground has been ceded to the alarmists" (Hockenbury and Upin 2012). "Winning the science", then, has nothing to do with participating in actual science or enhancing our understanding of the cause and effects of climate change. To win the science, rather, is to intentionally distort the degree of scientific understanding we have about AGW, so as not to "lose the moral high ground". The irony, of course, is that deniers employ morally nefarious tactics to do so.

We have already noted that there is virtually no uncertainty about the reality of AGW among actual climate scientists – and this was the case even in 1998. Suggesting otherwise, with the goal of influencing public opinion against scientific consensus, is a clear distorting influence. Moreover, this distortion exemplifies the malicious type of SEE-BS described above. Manufacturing and propagating viewpoints because they support one's political-economic ends, not because there are genuine challenges to "conventional wisdom", seeks to confuse rather than inform public opinion. That sowing this confusion is the denier's express goal highlights their malicious and exploitative intent.

These manufactured controversies can be notoriously difficult to detect. Often, the "controversy" is propagated by highly credentialed scientists, who clothe BS in actual science. Consider Heather Douglas's helpful description of *isolated fact* BS at the interface of climate science and policy (2006). This form of BS presents data that seems to contradict the conclusions offered by a scientific model, without considering how that data fits within the model holistically. This isolated data is then used to dismiss the model itself. An example is appeals to temperature drops between 1940 and 1975 that fail to take aerosols into account, but are used as "evidence" against CO_2-based AGW. Empirical data reveals that during a time of increased industrial expansion in the mid-twentieth century, the planet was cooling rather than warming. This data certainly offered a scientific puzzle. By the mid-1990s, however, climate scientists acknowledged that this puzzle was solved. As they discovered more about how aerosols impact global climate, they began incorporating them in their climate models. Once factored into these models, aerosols explained the cooling period. These are the same models that accurately predicted a warming period following the dissolution of atmospheric aerosols. Douglas notes that "including the fact of aerosols in one's understanding of climate records could be inconvenient, but ignoring aerosols produces bullshit ... Anyone who honestly participated in the climate change debate was aware of this crucial development" (Douglas 2006, 219). We have reason to conclude, then, that no matter how well credentialed, folks like S. Fred Singer are not *honestly* participating in the debate when supporting their claim that "the correlation between temperature and carbon dioxide levels is weak and inconclusive" (Singer 2008, 4) with the 1940–1975 cooling period.[7]

By utilizing various forms of BS, and having "expert" testimony supporting it, climate deniers willfully disseminate distorting influences seeking to stifle the public understanding of climate science. It is much simpler to convince people that there is indeed scientific uncertainty when two purported experts disagree and are each pointing to "scientific data" in support of their respective positions. There is no genuine disagreement, however, despite how things might appear to the lay person. Not all parties are communicating in good faith. An honest portrayal of the so-called debate would admit that actual practicing climate scientists accept the reality of AGW with near uniformity.

8.2.2 Distorting the Politics

In the late 2010s conditions were increasingly favorable for garnering public support *against* anyone accepting the scientific consensus surrounding AGW. We can note the overt appeal to partisan bias in Tim Phillips's admission that if "you win on the policy, you win on the science" (Hockenbury and Upin 2012). This tactic appeals to an already heightened partisan bias by associating climate change policy and the science it is based on with a specific political view – i.e., asserts that those who accept AGW are politically motivated to do so. Applied in this context, this highly effective tactic framed climate change policy as a political battle between so-called liberals and conservatives. Consider the release of the documentary *An Inconvenient Truth* in 2006 by former US Vice President and Democratic presidential candidate Al Gore, for example. This documentary's purpose was to raise public awareness of AGW and its potential impacts – a noble endeavor that ironically may have done more damage than otherwise. Indeed, James Taylor of the Heartland Institute makes the point unabashedly, asserting that Gore was "the best thing to happen for climate skeptics" (Hockenbury and Upin 2012). That a "liberal political elite" seemed to be leading the charge on climate change provided a ready-made antagonist for deniers to use as evidence of climate change as a political rather than scientific issue.

Further politicizing of climate change occurred when cap-and-trade legislation passed in the US Congress in 2009, but was subsequently blocked in the Senate in 2010. While many citizens were still reeling from the government's handling of the 2007–8 financial collapse, deniers painted these types of CO_2 emission-reduction policies as further unwarranted government intervention. When asked if the recession was an opportunity to push back against climate policy, Rep. James Susenbrenner (R) exulted: "It sure was". Utilizing partisan radio, television, political campaigns, and billboards, deniers rebranded climate change policy as a politically based rather than scientifically based issue. These propagandists employed hackneyed but effective BS, asserting that cap and trade was simply another "job killing and tax increasing" big-government policy that the taxpayers could not afford, for example.

By framing the issue as political rather than scientific, deniers were able to avoid the ratio problem concerning the more than 97% of climate scientists who accept AGW when pushing their desired conclusions. They were now able to skew the issue as a one-to-one partisan political disagreement, rather than a scientific one. Deniers were thus able to frame the consensus about AGW as consensus among "liberals" or "Democrats" and not as scientific consensus. As discussed above, appeals to partisan bias are extremely powerful persuasion tools (Mason 2018), especially when these biases are already established in connection to other partisan issues. Framing AGW's ontology as a political rather than scientific matter is one way to trigger extant partisan biases. Coupled with the distortion of both climate science and climate scientists' acceptance of AGW, this distortion technique significantly compromises one's ability to form rational judgments about the appropriate political response to AGW. How we should respond to AGW is a political issue, the reality of AGW is not. Arguing otherwise is BS. Arguing otherwise so that someone will act in a way that serves one's own end, while appearing to communicate in good faith, is a paradigm case of SEE.

8.3 SEE and Democratic Legitimacy

We have evidence that public opinion has little impact on policy decision making (Gilens and Page 2014).[8] It is clear from the above tactics, however, that those interested in preventing climate legislation and green energy initiatives are indeed interested in shaping public opinion. Consider again the 2010 US primary. The climate denial campaign was a catalyst for 10 to

15 seats in the US House of Representatives being flipped, simply because an incumbent accepted scientific consensus. A striking example of this phenomenon is six-term South Carolina congressional representative Bob Inglis's primary loss. Inglis has an exemplary conservative voting record, save for one issue: climate change. It was because he accepted the scientific consensus about the realities of climate change that Inglis was "primaried" by a climate denier who received 70% of the vote in the primary, and subsequently went on to win Inglis's congressional seat (Hockenbury and Upin 2012). This is a staggering example, revealing the effectiveness of SEE tactics. Fewer things will alter a politician's message than having that message turn politically toxic. And as Tim Phillips noted following the mid-terms: "If you buy into green energy ... you do so at your political peril" (ibid.).

Although the number of flipped seats might seem insignificant, the overall effect was significant. Consider that President Obama, throughout his 2007 presidential campaign and into the first two years of his administration, repeatedly acknowledged the importance of implementing climate change legislation. Concurrent with the denialist campaign in 2009, the Obama administration mentioned climate change 246 times – peaking in November before the Copenhagen summit. After this point, however, the administration began replacing "climate" rhetoric with "energy" rhetoric. By early 2011, on the heels of the mid-term elections, Obama's State of the Union only mentions climate change once in passing, where Obama all but admits that the issue's political toxicity makes meaningful discourse impossible; "energy", on the other hand, is mentioned 23 times in the same speech. We find similar trends occurring throughout the political sphere at that time, prompting the observation that "the phrases 'climate change' and 'global warming' become all but taboo on Capitol Hill, stunningly absent from the political arena from 2010 until the fall of 2012" (Kincaid and Roberts 2013, 45–46).

Of course, these effects did not end with simply changing political discourse. As documented in Hockenbury and Upin (2012), denialist policies were now able to be implemented with public support. Tennessee lawmakers mandated the teaching of climate denial in public schools.[9] In Virginia, the term "sea-level rise" was cut from all official documents, as it was deemed too partisan – another example of how the deniers were able to distort the political with the scientific. We have already noted that cap and trade, after being passed by the US House of Representatives, was blocked in the US Senate due to public backlash in senators' home ridings. Whether or not cap and trade is as effective as required given current climate projections, implementing the policy at least recognizes that there is a problem.

We have already noted some of the ethical implications of leaving CO_2 emissions unchecked and how addressing this crisis requires collective efforts. As our example highlights, however, such collective efforts in the context of democratic decision-making procedures are rendered inert when much of the electorate fails to recognize there is a problem to begin with.

This failure is symptomatic of an epistemically compromised electorate, where there are few shared epistemic standards for forming political judgments; rather, many members of the electorate base the rationality of their political judgments along partisan lines rather than epistemic ones. Our biases are then further calcified based on our epistemic interdependence. Our critical capacities concerning political judgments are stunted because of these biases and a misplaced trust in the information sources. Moreover, partisan divisions sow distrust among the electorate, preventing opportunities for meaningful political discourse.

Epistemic compromise is analogous to a compromised immune system. An immune system is reliable to the degree that it keeps one healthy and is unreliable to the degree it does not. The more compromised one's immune system, the more vulnerable one is to illness. Being epistemically compromised works in much the same way. The particular damage that partisan biases incur, for example, is not simply that they might result in irrational political judgments. Rather, it's the effect that these biases have on one's higher-order judgements, such as what qualifies as evidence for a given proposition, who one trusts as reliable testifiers, and the narrative one constructs about those who disagree with oneself. Moreover, closely associated with partisan bias is biased assimilation, at times referred to as the backfire effect, or double-down effect. The backfire effect occurs when one becomes more convinced a view is correct after being presented with evidence against that view. Rather than reason-responsive reactions to counterevidence, epistemic agents afflicted by assimilation bias hold fast, doubling-down their commitment to extant views. This latter bias prevents rational dialogue and epistemic reason-giving as a doxastic corrective, further evidencing the ill effects of epistemic compromise.[10]

As I have previously suggested (Olson 2015), these types of higher-order attitudes toward evidence make up one's doxastic, or belief-forming, disposition. We can understand epistemic compromise in relation to these dispositions in virtue-theoretic terms. The more epistemically compromised one is, the less disposed one is to exercise epistemic virtues; one who is epistemically compromised becomes more disposed to epistemic vices, such as dogmatism, rejection of expert opinion, accepting a proposition with insufficient evidence, or, worse, accepting a proposition in the face of contrary evidence. Each of these vices is symptomatic of partisan bias and related vices, such as group think, assimilation bias, and dogmatism. We can note how these vices are cancerous to one's set of judgments as they are self-reinforcing and will often frame future inquiry, further preventing one's ability for rational consideration.

When a sufficient percentage of an electorate is epistemically compromised in this way, I suggest the system is compromised. Political judgments resulting from an epistemically compromised citizen are more likely to be held irrationally, resulting in irrational actions – voting for candidates because they are part of "your team" despite most of their policies working against one's own best interests, for example. We thus observe that just as propositions accepted for epistemically vicious reasons are cancerous within one's set of judgments relevant to that proposition, so too are individual members of an electorate epistemically cancerous to the whole when they are epistemically compromised. Recall the overall effect that changing 15, of 435, congressional seats in the 2010 US mid-terms had on US political discourse surrounding climate change, described above.

8.3.1 Democratic Legitimacy

It is generally accepted that liberal democratic legitimacy rests on the need for shared political epistemic standards, often referred to as *public reason*. [11] Rawls offers an influential conception of this idea: "when constitutional essentials and questions of basic justice are at stake, the exercise of coercive power, the power of free and equal citizens as a collective body, is justifiable to all in terms of their free public reason" (2001, 141). Rawls elaborates:

> [the values of public reason] fall under the guidelines for public inquiry and for the steps to be taken to ensure that inquiry is free and public as well as informed and reasonable. They include not only the appropriate use of the fundamental concept of judgments, inference, and evidence, but also the virtues of reasonableness and fair-mindedness as shown in the adherence to criteria and procedures of commonsense knowledge and to the methods and conclusions of science when not controversial.
> (Rawls 2001, 91–92)

The coercive power members of an epistemically compromised electorate have over each other is not justified; public reason itself is subject to such compromise. A fortiori, the *free* exercise of public reason is inhibited by those epistemically exploitative practices undermining democratic legitimacy from the inside out. In our current example, those who exercised their coercive powers over others in the 2010 US mid-terms were themselves subject to more insidious coercive powers in the form of SEE. Not only did SEE succeed in undermining the political epistemic virtues of reasonableness and fair-mindedness, but also in undermining the adherence to uncontroversial science. Climate change policy is a question of basic justice, which any reasonable person should acknowledge; the science it is based on is uncontroversial.

One might counter that despite failing to meet the public reason ideal for shared epistemic standards, there are moral reasons supporting democratic legitimacy independent of any epistemic considerations. The procedural fairness in the democratic process, for example, might legitimize the coercive powers of the electorate. Insofar as the outcome of a democratic process results from a fair procedure, such as free elections for all eligible participants, the process is legitimate. Notice that the process itself, not the outcome of the process, determines the legitimacy of the procedure. Such *pure procedural* defenses of democratic decision making are subject to convincing counterarguments, however. A simple example will suffice.

Suppose A and B are considering which environmental policy to implement in order to mitigate climate change. Each has equal say in which policy is chosen, but they disagree on which policy to choose. Rather than appealing to the reasons favoring one policy over the other, however, they flip a coin because procedurally it is the fairest arbiter. In fact, A's policy choice is superior to B's for its stated goals. Assuming each genuinely desires that the policy is effective in achieving its purpose, flipping a coin to make this decision is irrational regardless of its procedural fairness.[12] That a procedure is maximally fair is insufficient justification for democratic decision making. An epistemic standard of rationality that increases the likelihood that the process produces the right decision seems necessary considering this objection.[13]

Given the necessity of epistemic considerations in conjunction with a Rawlsian conception of public reason, we require a political epistemology suitable for legitimizing democratic decision making. Elizabeth Anderson (2006) offers a helpful discussion here. Anderson contends that "the epistemic powers and needs of any institution should be assessed relative to the problems it needs to solve", which in a democracy include the "problems (a) of public interest, the efficient solution to which requires (b) the joint action by citizens, (c) through the law" (2006, 9). We judge the epistemic efficacy of democratic decision-making institutions predominantly, but only partially, by external factors: how well does the decision solve the problem of public interest. There are internal factors of assessment as well, however, namely,

> whether a problem counts as of genuine public interest is determined in part by whether it is an actual object of public concern – that is by whether citizens or their representatives affirm its place in the public agenda through procedurally fair decision-making processes.
>
> (Anderson 2006, 10)

From these epistemic needs, Anderson proposes that Dewey's *experimentalist* account of democracy provides the best model for democratic epistemology.[14] This model emphasizes the cooperative powers of social groups to solve practical problems, which Anderson describes as "cooperative social experimentation ... the application of the scientific method to practical problems ... abandoning dogmatism, affirming fallibilism, and accepting the observed consequences of our actions as the key evidence of prompting us to revise them" (2006, 11; rearranged).

Combining Anderson's proposal to contextualize an electorate's epistemic powers to its democratic needs with Rawls's guidelines for public reason, we have a general template for assessing both the epistemic success of the democratic process and where improvements are needed when success is limited. Assessing the epistemic powers of US democracy in 2010, we find that Anderson's three proposed needs are present. Public interest in climate change policy was sufficient to manifest joint legal action through the institution of voting, which successfully eliminated cap and trade as well as political discourse defending climate change's reality. Insofar as the prevention of implementing policies to mitigate CO_2 emissions is what the electorate desired, the voters "got it right". It is highly plausible that many of these voters were highly informed through traditional sources of information – television, radio, and print media; their political representatives; and testimony from their friends and family. As discussed above, however, much of this information served as a distorting influence concerning the underlying facts to which the policies were responding – viz the reality of AGW. Were voters working with reliable information, it is unclear that they would have desired this outcome. In this latter sense, we cannot assess the outcome of this process as actually successful. The electorate was epistemically compromised.

An implication, and virtue, of Anderson's epistemic experimentalist defense of democracy is that when functioning properly, it promotes self-correction. If a political decision fails to achieve its stated ends, then through deliberation, voting, and referendums, that decision can be amended or overturned. Accepting that the 2010 US mid-term results achieved the electorate's stated ends, but that it is unclear that these would be their ends had the electorate acknowledged the irrationality of not accepting scientific consensus, we might judge this structure in need of self-correction. Notice, however, that an epistemically compromised electorate's ability to self-correct *in this way* is significantly hamstrung by SEE's distorting influence. The shared standard of accepting expert scientific consensus is what those employing SEE in our current example seek to undermine. Self-correction under these circumstances is unlikely. If this intuition is accurate, we find that an epistemically compromised electorate, in addition to compromising the shared standards necessary for public reason, also compromises an electorate's epistemic ability to self-correct. We again find, accepting Anderson's promising epistemic defense of democratic legitimacy, that an electorate's epistemic compromise undermines the democratic decision's legitimacy.

I should like to emphasize that I am not suggesting that members of the electorate lack the epistemic capabilities to make rational political decisions in general. I am suggesting, however, that members of an epistemically compromised electorate have significantly diminished political epistemic reliability due to accepting and employing intentionally distorted information when forming political judgments. Moreover, it is feasible that many members of the electorate are epistemically blameless, despite being epistemically compromised, as they are unknowingly subjected to SEE – recall that our epistemic interdependence and partisan biases are systematically unavoidable. Nevertheless, under SEE circumstances, an electorate can be sufficiently unreliable to call into question its democratic legitimacy. Furthermore, perhaps the strongest case for democratic legitimacy, viz

Anderson's experimentalist epistemic defense, can neither retain nor regain democratic legitimacy when an electorate is epistemically compromised.

8.4 Mitigating SEE in Liberal Institutions

I suggest that the forgoing considerations support mitigating the exploitative and willful dissemination of distorting influences from those liberal institutions and practices informing the electorate on matters of political and public interest. On the plausible assumption that we should be morally concerned with maintaining democratic legitimacy, it is in an electorate's best interest to have a shared set of political epistemic values from which they can form rational political judgments; we should be concerned with public reason. As noted above, symptomatic of an epistemically compromised electorate is a failure to maintain shared rational standards when forming political judgments. Expert scientific consensus, for example, is one rational standard for evaluating claims about the natural world. If distorting influences result in 50% of an electorate accepting the reality of AGW with 50% rejecting expert consensus, then we have grounds for judging the electorate epistemically compromised. The legitimacy of democratic outputs and their potential for self-correction is called into question under these conditions. We should therefore seek to prevent epistemic compromise among an electorate.

Of specific concern in our current context is SEE. As I have described it, SEE manifests systematically within certain democratic structures and institutions. Given its scope and scale, the challenges SEE presents far exceed any potential mitigating solutions one could offer in a single chapter. Nevertheless, I suggest that coupling a democratic citizen's epistemic needs with those institutions whose express purpose is the facilitation of public reason reveals the need for reforms in some of these institutions. Consider journalism, for example – hereafter, the press.[15] In a society's necessary division of epistemic labor, the press's role is to keep its audience informed of the most pressing issues they face. A fortiori, a liberal democratic conception of the press should jibe with the shared epistemic needs of its target audience. As we are all members of communities and countries, and with the impacts of globalization, there is an epistemic need for reliable local, federal, and international news sources. It is with this conception of the press and specific level of generality to which it applies that we should assess how well it fulfills this purpose. We find, however, that these expectations are often unmet.

What is worse, institutions like journalism are often participants in SEE. Consider the distorting influence found in partisan slanting. As above, partisan bias is a substantial epistemic vulnerability, made even stronger when coupled with our epistemic interdependence. Despite this fact, blatant appeals to partisan bias on various broadcast news venues are not uncommon. These appeals are clear cases of SEE. The reporting provided by these sources is only peripherally related to clear and accurate information concerned with viewers' epistemic needs. Rather, the organizations' foremost concerns are maintaining viewers for advertising revenue, which is well served by base pandering. The existence of these news sources has a dual compromising effect. First, they encourage confirmation bias and selective inquiry by pandering to their viewers. Second, according to Mason (2018), appeals to partisan bias have powerful compromising effects when employed to manufacture public opinion along partisan lines.[16] Suppose, for example, that a news network is known as the conservative, or in our current example, Republican, source for news. If this news source consistently reports that climate change is not established science, they are informing conservatives of what their position should be qua conservatives. If most viewers have no opinion on the reality of climate change before observing the news, they will most likely have one afterwards. Partisan bias in those news organizations pandering to so-called liberals is equally compromising. Suppose these organizations accurately report the climate science and its ramifications, while also pandering to their viewers by uncritically echoing unsubstantiated Democrat talking points.[17] For someone associating as a conservative, this pandering provides evidence that climate change is a liberal, rather than scientific, issue. The epistemically compromising effect of a partisan press fails its mandate to fill an electorate's epistemic needs. The institution thus loses its legitimacy.

Space forbids a full analysis of the shared political epistemic standards the press should exhibit and how we might institute and enforce such standards. We can consider avenues for further discussion along these lines, however. Recent works by Sandford Goldberg, for example, provide promising discussions on social epistemic obligations and expectations that we might apply here. Goldberg (2017) proposes that when one fails to meet certain context-sensitive social epistemic obligations when forming a judgment, that judgment is defeated, thereby failing the justification norm of assertion.[18] Extending this idea further, Goldberg (2018) advances additional normative constraints on our epistemic justification, given context-sensitive social epistemic expectations we reasonably place on each other. We can expect, for example, that an anesthesiologist knows the safe amount of anesthetic to administer in a specific context – that is something the anesthesiologist *should* know. Coupling Goldberg's social epistemic normative framework with Anderson's epistemic conception of democracy and the press's role of providing for its viewer's epistemic needs, we can contextualize these normative constraints to institutions such as the press.

Returning to our original example, if a news source is reporting on climate change, it *must* acknowledge the consensus among active climate scientists that AGW due to CO_2 emissions is our reality. A failure to acknowledge this point is either misleading or epistemically irresponsible. It is misleading because the scientific consensus surrounding AGW is ignored, which contributes to the audience misunderstanding information relevant to their lives – i.e., AGW's reality. Borrowing Douglas's apt turn of phrase, reporting this fact might be inconvenient, but ignoring it is BS. This form of BS, as discussed above, is indefensible within the social moral epistemic parameters of liberal democracy. Failing to acknowledge scientific consensus is also epistemically irresponsible. Appeals to ignorance are indefensible given the press's role in fulfilling its audience's epistemic needs and our reasonable epistemic expectations of the press. The scientific consensus concerning AGW is information a news organization *should* know about.[19] Anyone seeking to challenge that consensus should not be given an official platform without having to respond to that fact. If there is a science-based policy initiative that will impact an electorate, then it is the news organization's epistemic responsibility to report the science accurately. A failure to report this information is either willfully misleading, or epistemically improper given our expectations. In either case, not reporting this information serves as a distorting influence; if the failure to report is motivated by an organization's self-interest, then it is a case of SEE.

Pragmatic solutions to challenges of the above type need not be drastic. Journalistic oversight is already in place in many countries. Moreover, those institutions overseeing journalism and broadcast standards often recognize the importance of epistemic standards. In the US, for example, the Federal Communications Commission acknowledges that "broadcasters may not intentionally distort the news" and that "rigging or slanting the news is a most heinous act against the public interest" (fcc.gov). Despite this recognition, the FCC is prohibited by law from infringing on broadcasters' rights to free speech. Notice, however, that there are certain restrictions on freedom of speech and expression that are uncontroversial. We are legally prohibited from expressing defamation, slander, libel, and perjury. These limitations on speech are justified because of the harm they cause targets of defamation or, in the case of perjury, to the system these freedoms are mandated to uphold. SEE statements are equally, perhaps even more, harmful than these other forms of outlawed speech. On pains of consistency, those harms justifying the suppression of perjury, for example, also justify suppressing SEE in specific liberal institutions. Maintaining and enforcing broadcast journalistic standards already recognized by institutions like the FCC offers the type of solution that could help to curb cases of SEE.[20] These shared standards should prevent, e.g., creeping partisan slanting, as implicit in these standards is what information is important for its audience; it's *that* information that should be presented, without misplaced emphasis or misleading connotations in its presentation. Failing to meet these epistemic journalistic standards would be subject to censure and could prompt litigation. Similar considerations apply *mutatis mutandis* to additional institutions, such as public education and political campaigning.

8.5 Conclusion

Using the United States as an example of a contemporary liberal democracy, I considered how public opinion on climate change affected its 2010 mid-term election. I argue that this example reveals how an electorate can become epistemically compromised through social epistemic exploitation. This form of exploitation preys on social epistemic vulnerabilities, such as our epistemic interdependence and our individual partisan biases. SEE occurs through the willful dissemination of distorting influences, under the pretense of communicating in good faith, with the goal of one's audience forming judgments that will manifest in actions to one's benefit. There is a social moral epistemic argument in favor of inhibiting this type of exploitation, as it undermines democratic legitimacy. This conclusion suggests we re-establish social epistemic parameters of our traditional liberal institutions whose mandate is to enhance public reason, updated to current epistemic needs – in our current case a set of shared epistemic standards that prevent social epistemic exploitation found in our liberal institutions and practices necessary for public reason.

Acknowledgements

Thanks to the participants at the Hungarian Academy of Science's "Science, Freedom, Democracy" workshop, the editors, and anonymous referees for very helpful feedback.

Notes

1 Here is a small sample of the global breadth of AGW and the cost to human life resulting from extreme heat over the past decade: 2010: 14,000 deaths in Moscow; 2013: 765 death in the UK; 2014: 1,877 deaths in Argentina; 2015: over 2,500 deaths in India and 3,275 deaths in France; 2018: 70 deaths in Canada (EM-DAT). Heat deaths are diminishing as those places most affected by rising temperatures have begun adapting, not because they are cooling.
2 A strong case can be made that this moral issue is deeply intertwined with neo-liberal economics and industrial globalization, which suggests that there will be no energy reform without economic reform (Klein 2014).
3 Social moral epistemology is introduced by Allen Buchanan as "the study of the social practices and institutions that promote (or impede) the formation, preservation, and transmission of true beliefs so far as true beliefs facilitate right action or reduce incidence of wrong actions" (2002, 133).
4 A distorting influence can be anything that inhibits one's ability to rationally consider a proposition, such as lack of sleep, intoxication, or false testimony. For present purposes, when referring to *distorting influences*, I mean *willfully propagated testimonial* distorting influences.
5 Mason argues that to some extent all politics are identity based, but not in the way that identity politics is often currently understood, viz, how one's individual social identities, like race, influence one's political behavior. Rather, the partisan political group with which one most strongly identifies creates a "sorting effect", wherein our social identities coalesce and homogenize around our political, or partisan, identity. One does not simply identify as a Republican, but most likely also as a Christian, a conservative, a climate-change denier, and so forth.
6 See Naomi Oreskes and Erik M. Conway's important work (2010) highlighting the ironic group of politically and economically motivated scientists for hire who

rejected the connection between cigarette smoke to lung cancer, the damaging effects DDT has on ozone, the contribution of air pollution to acid rain, and most recently the science supporting anthropogenic global warming.

7 We are given further evidence of the bad-faith BS coming from deniers like Singer in his (2010) response to a literature review on the climate change debate by Philip Kitcher (2010). Singer accuses Kitcher of "clearly not understanding climate science" (Singer 2010, 849), but offers nothing to clarify where Kitcher has erred or what science Kitcher should consider. Rather, Singer ironically proceeds with a series of ad hominem, red herring, and tu quoque fallacies throughout his rebuttal. How this response survived peer review, or even editorial approval, is mysterious.

8 Gilens and Page highlight that in the US, neither individual citizens nor mass-based interest groups have any independent influence on legislation. Conversely, financial elites and lobbyists representing business interests have "substantial independent impacts on US government policy" (2014, 564).

9 This echoes the mantra "teach the controversy", itself a form of BS, concerning evolution versus creationism curricular disputes.

10 This is one example of the way partisan bias exacerbates the ill effects of other biases. There are many more examples; see McRaney (2011; 2013).

11 There are notable critics, however. For a pragmatist rejection of public reason as the by-product of scientific and philosophical modernity, see Gutting (1999). For a challenge to consensus as a democratic virtue, see Rescher (1993).

12 This is a variant on an example presented in Anderson (2008) and Estlund (2008b, chapter 1).

13 There is a literature arguing the merits of epistocracy, which would restrict suffrage to individuals who are epistemically most able to make the correct decisions. See, e.g., Brennan (2011), Mulligan (2015), and Ancell (2017) for helpful discussions on this issue. As the focus of this chapter is concerned with extant liberal democracies, non-democratic theories are beyond its scope.

14 Two other popular democratic epistemologies are the well-known Condorcet Jury Theorem and the Diversity Trumps Ability Theorem; see Hong and Page (2004). See Estlund (2008a) for an overview of additional varieties of epistemic democracy.

15 In what follows, I provide the type of reflection I should like to emerge from the foregoing conclusions but am not in the details committed to the proposals made in the remainder of this section. These proposals merely serve as an example of the types of discussions we should have about those specific liberal institutions whose purpose is to facilitate public reason, such as news, education, and political campaigning. My hope is that those who are best placed to offer solutions to the social moral epistemic challenges discussed above, accept my argument that it is imperative we do provide such solutions.

16 Noam Chomsky and Edward Herman (2002) provide convincing evidence of this phenomenon's ubiquity in American corporate media outlets.

17 Incessant reporting on Trump-Russia relations from 2016–19 provides a clear example of this phenomenon.

18 Full disclosure: I disagree with an important contention in Goldberg's position, which is that unpossessed evidence can defeat an individual's epistemic justification – what Goldberg refers to as a normative defeater. I nevertheless find his discussions on social epistemic obligations very insightful and endorse something akin to his view, when contextualized to liberal institutions such as the press.

19 For Goldberg, this *should* is epistemic, as a failure to know what one should defeats one's ability to know. I would contextualize this normative assessment as a social moral epistemic *should*, where the epistemic ought is also a moral ought.

20 Legal parameters concerning epistemic obligations are not without precedent. Medical professionals are legally expected to know and disclose various medical treatments. Lawyers legally should understand legal precedent. An electrician is held legally responsible for keeping her work up to code. And so on.

Bibliography

American Petroleum Institute. 1998. "Global Climate Science Communications Plan." Available online at insideclimatenews.org/documents/global-climate-science-comm unications-plan-1998. Accessed May 30, 2020.

Ancell, A. 2017. "Democracy Isn't that Smart (But We Can Make it Smarter): On Landemore's Democratic Reason." *Episteme* 14 (2): 161–175.

Anderson, E. 2006. "The Epistemology of Democracy." *Episteme* 3 (1–2): 8–22.

Anderson, E. 2008. "An Epistemic Defence of Democracy." *Episteme* 5 (1): 129–139.

Brennan, J. 2011. "The Right to a Competent Electorate." *Philosophical Quarterly* 61 (245): 700–724.

Buchanan, A. 2002. "Social Moral Epistemology." *Social Philosophy and Policy* 19 (2): 126–152.

Chomsky, N., and E.S. Herman. 2002. *Manufacturing Consent: The Political Economy of Mass Media*. New York: Pantheon Books.

Cook, J.*et al.*2013. "Quantifying the Consensus on Anthropogenic Global Warming in the Scientific Literature." *Environmental Research Letters* 8: 024024. Available online at iopscience.iop.org/article/10.1088/1748-9326/8/2/024024/pdf. Accessed June 13, 2020.

Dewey, J. 2010. *Teachers, Leaders, and Schools: Essays by John Dewey*, edited by D. J. Simpson and S.E.Stack, Jr.Carbondale and Edwardsville: Southern Illinois University Press.

Douglas, H. 2006. "Bullshit at the Interface of Science and Policy: Global Warming, Toxic Substances, and Other Pesky Problems." In *Bullshit and Philosophy*, edited by G.L. Hardcastle and G.A. Reisch, pp. 215–228. Chicago: Open Court.

Estlund, D. 2008a. "Introduction: Epistemic Approaches to Democracy." *Episteme* 5 (1): 1–4.

Estlund, D. 2008b. *Democratic Authority: A Philosophical Framework*. Princeton, NJ: Princeton University Press.

Federal Communications Commission. N.d. "Complaints About Broadcast Journalism." Available online at fcc.gov/consumers/guides/complaints-about-broadca st-journalism. Accessed June 12, 2020.

Foote, E. 1856. "Circumstances Affecting the Heat of the Sun's Rays." *American Journal of Science and Arts* Second Series 31: 382–383.

Frankfurt, H. 2005. *On Bullshit*. Princeton, NJ: Princeton University Press.

Gilens, M., and B.I. Page. 2014. "Testing Theories of America Politics: Elites, Interest Groups, and Average Citizens." *Perspectives on Politics* 12 (3): 564–568.

Goldberg, S. 2017. "Should Have Known." *Synthese* 194: 2863–2894.

Goldberg, S. 2018. *To the Best of Our Knowledge*. Oxford: Oxford University Press.

Gutting, G. 1999. *Pragmatic Liberalism and the Critique of Modernity*. New York: Cambridge University Press.

Hansen, J. 1988. "Statement of Dr. James Hansen, Director, NASA Goddard Institute for Space Studies." In *United States Senate Committee on Energy and Natural Resources: Greenhouse Effect and Global Climate Change: Hearings Before the*

Committee on Energy and Natural Resources, pp. 36–46. United States Senate, One Hundredth Congress, First Session. Washington: US Government Printing Office.

Hilton, I. 2008. "The Reality of Global Warming: Catastrophes Dimly Seen." *World Policy Journal* 25 (1): 1–8.

Hockenbury, J., and C. Upin. 2012. "Climate of Doubt." *PBS Frontline*, S30 E20, originally aired October 22, 2012. Available online at pbs.org/video/frontline-clima te-of-doubt/.

Hong, L., and S.E. Page. 2004. "Groups of Diverse Problem Solvers Can Outperform Groups of High-Ability Problem Solvers." *Proceedings of the National Academy of Sciences* 101: 16385–16389.

Kincaid, G., and J.T. Roberts. 2013. "No Talk and Some Walk: Obama Adminis-tration First-Term Rhetoric on Climate Change and US International Climate Budget Commitments." *Global Environmental Politics* 13 (4): 41–60.

Kitcher, P. 2010. "The Climate Change Debates." *Science* 328: 1230–1234.

Klein, N. 2014. *This Changes Everything*. Toronto: Knopf Canada.

Leiserowitz, A., E. Maibach, S. Rosenthal, J. Kotcher, M. Ballew, M. Goldberg, and A. Gustafson. 2018. *Climate Change in the American Mind: December 2018*. New Haven, CT: Yale Program on Climate Change Communication.

Lindsey, R. 2020. "2019 was Second-Warmest Year on Record." NOAA. Available online at www.climate.gov/news-features/featured-images/2019-was-second-warm est-year-record. Accessed January 17, 2020.

Mason, L. 2018. *Uncivil Agreement: How Politics Became Our Identity*. Chicago: University of Chicago Press.

McRaney, D. 2011. *You Are Not So Smart*. New York: Gotham Books.

McRaney, D. 2013. *You Are Now Less Dumb*. New York: Gotham Books.

Mulligan, T. 2015. "On the Compatibility of Epistocracy and Public Reason." *Social Theory and Practice* 41 (3): 458–476.

Olson, D. 2015. "A Case for Epistemic Agency." *Logos and Episteme* 6 (4): 449–474.

Oreskes, N., and E.M. Conway. 2010. *Merchants of Doubt: How a Handful of Sci-entists Obscured the Truth on Issues from Tobacco Smoke to Global Warming*. New York: Bloomsbury Press.

Rawls, J. 1971. *A Theory of Justice*. Cambridge, MA: Harvard University Press.

Rawls, J. 2001. *Justice as Fairness: A Restatement*. London: Harvard University Press.

Rescher, N. 1993. *Pluralism: Against the Demand for Consensus*. New York: Oxford University Press.

Singer, S.F. (ed.). 2008. *Nature, Not Human Activity, Rules the Climate. Summary for Policymakers of the Report of the Nongovernmental International Panel on Climate Change*. Chicago: The Heartland Institute.

Singer, S.F. 2010. "A Response to 'The Climate Change Debates'." *Energy and Environment* 21 (7): 847–851.

Tyndale, J. 1861. "The Bakerian Lecture: On the Absorption and Radiation of Heat by Gases and Vapours, and on the Physical Connexion of Radiation, Absorption, and Conduction." *Philosophical Transactions of the Royal Society of London* 151: 1–36.

UN Environment Program. 2020. "Ten Impacts of the Australian Bushfires." Avail-able online at www.unenvironment.org/news-and-stories/story/ten-impacts-austra lian-bushfires. Accessed January 22, 2020.

Part III

Freedom and Pluralism in Scientific Methodology and Values

9 Are Transparency and Representativeness of Values Hampering Scientific Pluralism?

Jeroen Van Bouwel

FACULTY OF ARTS AND PHILOSOPHY, GHENT UNIVERSITY

9.1 Introduction

It seems to be increasingly accepted among philosophers of science that values influence the scientific process, unavoidably.[1] In his excellent book, *A Tapestry of Values*, Kevin Elliott (2017) demonstrates convincingly, using a lot of lively examples, how science is permeated with economic, cultural, ethical, and political value judgments. Elliott (2017, 8–9) distinguishes five avenues for value influence that can be connected to five questions regularly shaping scientific research, namely:

a   What research topics to prioritize?
b   What specific questions to raise, methods to use, assumptions to make?
c   What are the aims of inquiry in this particular context (weighing a variety of theoretical and practical goals, e.g., a quick or inexpensive fix rather than a slower, more detailed result)?
d   How to deal with questions of uncertainty (e.g., when is the available evidence sufficient for particular sorts of conclusions)?
e   How to report and frame the conclusions of scientific research (e.g., what terminology, categories, or metaphors to employ in providing scientific information)?

Doing science implies answering these questions, making value judgments, be it implicitly or explicitly. In this chapter, I consider the presence of these value influences in science as a given, as the starting point.[2] The next question is: under what conditions is the influence of these values in science justifiable? Elliott (2017, 10) stipulates three conditions that "appear to be particularly important for bringing values into science in an appropriate fashion" and "why some influences appear to be more justifiable than others" (ibid.): value influences should be (1) made as *transparent* as possible, (2) *representative* of our major social and ethical priorities, and (3) scrutinized through *engagement* between different stakeholders, engaging "a wide variety of stakeholders – citizen groups, policymakers, scientists, and other scholars – to identify value-laden aspects of science and to reflect on how to address them" (Elliott 2017, 133).

The aim of this chapter is to scrutinize Elliott's conditions (1) and (2). This scrutiny is done along considerations of scientific pluralism and productive social-epistemic constellations in connection with recent evolutions in political science and in economics. I suggest some modifications to Elliott's understanding of the transparency and representativeness conditions and I show how they can be fulfilled committing to both epistemic productivity and democracy. Let us start with scrutinizing the first condition, transparency.

9.2 Scrutinizing Elliott's First Condition: Transparency

9.2.1 The Transparency Condition in General

Elliott's first condition, *transparency*, is a frequently mentioned condition or requirement when considering how to deal with values in science. The general idea is for scientists to be as clear as possible about their "data, methods, models, and assumptions so that others can identify the ways in which their work supports or is influenced by particular values" (Elliott 2017, 14). Transparency obviously has benefits, contributing to the credibility and legitimacy of knowledge claims (cf. Lupia and Elman 2014, 20). However, the transparency requirement also raises some general questions:

a   *How much information do we have to share about our choices and judgments?* Would we not all too easily be submerged by too much information if all choices and judgments were being made transparent? There might be cases in which the transparency requirement can be fulfilled smoothly with concise information. However, when we think, for instance, of political scientists measuring democracy or economists calculating GDP, these calculations will imply many choices and judgments. Explaining one particular value choice might already be hard, not to speak of enumerating the alternative choices one might have opted for. What if we combine all of these choices and possible interactions between them?

b   *To what extent can you make value judgments transparent?* Are value choices not engrained in our scientific tools, in the history and path dependency of disciplines (choices made a long time ago that pushed the discipline in a certain direction and closed off possible alternative tracks)? How deep and how far back can we go? This leads us to the following question.

c   *To what extent are scientists aware of their own values, value judgments, and background assumptions?* Many history of science studies show how scientists projected values typical of their historical context into science, most likely being unaware of it. This skepticism about transparency is not new. The philosopher Ernest Nagel wrote in 1961:

> Although the recommendation that social scientists make fully explicit their value commitments is undoubtedly salutary, and can produce excellent fruit, it verges on being a counsel of perfection. For the most

part we are unaware of many assumptions that enter into our analyses and actions, so that despite resolute efforts to make our preconceptions explicit some decisive ones may not even occur to us. But in any event, the difficulties generated for scientific inquiry by unconscious bias and tacit value orientations are rarely overcome by devout resolutions to eliminate bias. They are overcome, often only gradually, through the self-corrective mechanisms of science as social enterprise.

(Nagel 1961, 489)

In the last sentence, Nagel suggests a more social-epistemic take on value influences, a track I will follow in Section 9.3 and following.

d Does more transparency lead to more/less trust in science? Besides raising questions about the extent to which scientists can and should make value judgments transparent, one can also wonder how transparency about values would affect the standing of science with the public. Some empirical studies actually seem to show that transparency about values reduces the perceived credibility of scientists. One study by Kevin Elliott himself, with co-authors, examined how citizens look at scientists who publicly acknowledge values. The authors conclude that their

results provide at least preliminary evidence that acknowledging values may reduce the perceived credibility of scientists within the general public, but this effect differs depending on whether scientists and citizens share values, whether scientists draw conclusions that run contrary to their values, and whether scientists make policy recommendations.

(Elliott et al. 2017, 1; for similar findings, see Yamamoto 2012)

However, it seems to me that this should not immediately be reason for despair among defenders of transparency requirements. The reduction in perceived credibility might be a consequence of a public being brought up with an ideal of value-free science (value-free understood as free from so-called non-epistemic values). This public perception could be shifted by a learning process, by reading Elliott's (2017) book, for instance, adjusting the image of science as value-free by acknowledging the unavoidable influence of values in science.

e Might a transparency requirement needlessly restrict scientists' methodological options or skew the scientific community? Besides looking at the possibilities of being transparent and what the consequences on the public's trust in science are, one can also ask what the impact of a transparency requirement on the scientific community as a whole might be and on the distribution of options and opportunities among the members of that community. I discuss the social-epistemic constellation of the scientific community in more general terms in Section 9.3. Let us first address this question by examining a recent debate in political science.

9.2.2 The Transparency Condition in Political Science

9.2.2.1 The Introduction of the Transparency Condition in Political Science

In October 2012 the American Political Science Association (APSA) amended its *Guide to Professional Ethics in Political Science* following the suggestions of the *Data Access – Research Transparency* (DA-RT) initiative. Section 9.6 of the Ethics Guide of APSA became: "Researchers have an ethical obligation to facilitate the evaluation of their evidence-based knowledge claims through data access, production transparency, and analytic transparency so that their work can be tested or replicated" (APSA 2012, 9). In the subsections 9.6.1 to 9.6.3 of the Guide one finds some further specifications about what *data access, production transparency*, and *analytic transparency* exactly mean and the formalities it implies.

In 2014, the DA-RT initiative led to the "Journal Editors' Transparency Statement" (JETS) (Elman, Kapiszewski, and Lupia 2018). This statement was signed by more than 20 editors of political science journals who pledged to commit their journals to the principles of access and transparency. In some quarters, JETS was received with a lot of skepticism and distrust, which cannot be a surprise given the central role journal publications play in the discipline – be it in allocating research funding, appointing faculty, ranking departments, and so on (see, e.g., Hicks et al. 2015 for more on the role of journals and bibliometrics in research evaluation and funding).

In reaction, the Qualitative and Multi-Method Research (QMMR) section of the APSA started up a participatory Qualitative Transparency Deliberations (QTD) process in 2015. The QTD process is driven by the concern that journals implementing the new data access and research transparency (DA-RT) requirements risks being incompatible with the plurality of logics of inquiry in the field of political science.[3] Therefore, the benefits and costs of the DA-RT transparency requirements, especially for the qualitative empirical approaches in political science, have to be scrutinized.

9.2.2.2 The Reception of Transparency Requirements in Political Science

Unsurprisingly, the idea of transparency requirements has been received differently depending on the research approach one is working in. This is illustrated in the report written by Kreuzer and Parsons (2018) as part of the QTD. They distinguish five research approaches (or traditions, as they call them), namely the (1) frequentist/experimentalist tradition, (2) Bayesian/process-tracing tradition, (3) comparative historical tradition, (4) modern constructivist tradition, and (5) interpretivist tradition. For each approach, *transparency* can be more or less problematic and/or filled in differently. The further you go down the list of traditions, from (1) to (5), the more skeptical researchers become with respect to the transparency requirements as proposed by the DA-RT initiative.

Let me here just mention some characteristics and worries of these different research traditions with respect to transparency requirements as presented by Kreuzer and Parsons (2018) in order to make my social-epistemological points. For a more comprehensive characterization of the traditions and their reception of the transparency requirements, I refer the reader to the full report.

1 The frequentist/experimentalist tradition consists of experimentalists and proponents of statistical analysis who conceptualize the social world as analogous to the physical world and within which evidence is contextless and lacks any specific historical coordinates. This tradition gets along fine with the DA-RT requirements – these requirements closely reflect the methodology of this tradition.

2 The Bayesian/process-tracing tradition has a more conditional and contextual view on knowledge production than (1). The interpretation of evidence depends on an analysis of preceding knowledge claims, the consideration of alternative explanations as well as of the concrete context of evidence. Therefore, it has a more encompassing understanding of transparency than (1), being transparent about the additional conditional and contextual dimensions. However, all of this is still well in line with the DA-RT requirements.

3 The comparative historical tradition has a bit more complex temporal sequencing compared to the former two traditions and also pays more attention to context. Researchers in this tradition are often seeking a whole chain of evidence that must support a hypothesis, with transparency practices involving a series of informal judgments elaborated in detailed historiographies, extended footnotes, or lengthy book reviews. Scholars in this tradition tend to endorse DA-RT requirements – the chain of evidence must be observable – but they worry about its practical limits; making all steps taken transparent (starting from far back into the research design) may be cumbersome in practice and contribute little to compelling results.

4 The modern constructivist tradition, just like tradition (5) below, considers human action to be operating through interpretive social constructs, be it ideas, norms, beliefs, identities, cultures, etc. These constructs are then considered as interpretive, human-made ideational filters through which people perceive themselves, their surroundings, and shape their actions. Accordingly, scholars in the modern constructivist tradition focus on thick evidence of discourse and action and pay detailed attention to the meaning of evidence in its precise social context. For scholars studying politics in these socially constructed contexts "the connotation of transparency – seeing through obstructions to reveal reality – suggests a misleading confidence in objective process and results" (Kreuzer and Parsons 2018, 10).

5 The interpretivist tradition shares the emphasis on social construction and, moreover, acknowledges that scholars too only access the world through social constructs, rejecting correspondence theories of truth, instead constructing coherent narratives. Here, as well, evidence is characterized as thick, detailed, taking into account socially constructed contexts. One understands that the notion of making available "raw data", as present in DA-RT, strikes interpretivists as odd: in their view, data never have the autonomous status implied by "rawness", and should be read in holistic, relational, and/or intersubjective context as much as possible. In general, Kreuzer and Parsons conclude that

> It surely makes sense for interpretivists to reject the DA-RT vision of transparency, because a core point of the interpretive tradition is that we cannot ever "see" through our socially-constructed filters. When scholars rooted in the frequentist/experimental tradition exhort inter-pretivists to articulate "transparency standards", interpretivists perceive a "tyranny of light" that is at best naive, and more likely to be a move to delegitimize their work.
>
> (Kreuzer and Parsons 2018, 11)

Thus, having briefly reviewed the reception of DA-RT among different research traditions, one has to acknowledge that there is considerable dis-content concerning and opposition to DA-RT principles, in particular with respect to their inappropriateness for some qualitative research methods. Let us use this DA-RT case in political science to formulate some concerns about the transparency condition.

9.2.3 Concerns about the Transparency Condition

a Proponents of the DA-RT initiative label the initiative as "epistemologically neutral" (cf. Lupia and Elman 2014, 20), i.e., not in any way uniformizing or affecting the existing epistemological differences within political science. This seems hard to defend, given the different standpoints among research traditions vis-à-vis the transparency condition I discussed above. As the discussion of the reception among research traditions also illustrates, the transparency require-ments rather risk intentionally or unintentionally discounting certain established approaches as being science or good scholarship. Rather than neutral, it might be skewing the scientific community as well as *hampering scientific pluralism* (cf. Section 9.3).[4]

b In combination with JETS, the DA-RT initiative might have serious consequences. Once journals subscribe to the transparency requirements as formulated in the APSA Ethics Guide, some methods might lose their status, because important journals stop publishing studies that use those methods.[5] Monroe (2018), for instance, illustrates this with her own work based on nar-rative interviews. Thus, methods might become underdeveloped, lower in the hierarchy, neglected. (This might affect both research and teaching, as social scientists teach some but not all methods to their students.) Thus, methodolo-gical options might become more restricted, which also influences the choice of research questions, or even what can be questioned. Reducing methodological choices might then "hinder political science's ability to address important political questions" (Monroe 2018, 142) as well as make "an implicit value judgement about which questions are important to be addressed" (Reiss 2017, 145). This scenario with transparency requirements advocated by JETS – *implemented into journals – restricting methods – not able to address certain important questions*, is an example of what I call *hampering scientific pluralism*.

   c   These questions and concerns about transparency have led to pleas for replacing research transparency by less demanding conditions. Some suggest *openness* or *explicitness*. Kreuzer and Parsons write in their report:

> It therefore makes sense to substitute research transparency with a more expansive, less formalized term like research explicitness. Especially prominent in posts on our group's forum … is the interpretive tradition. It rejects the notion of research transparency, but may be open to explicating their research methods to be consistent with other ways of critically evaluating scholarly interpretations and explanations.
>
> (Kreuzer and Parsons 2018, 1)

The general idea seems to be that it is important to have scientists clarify and discuss the important choices they make and the values that are part of their practice (going beyond the value-free ideal), but that transparency is too demanding, unnecessary, or utopian an aim.

   Within the QTD process, there is also a group that proposes to develop "Community Transparency Statements" (CTSs) as an alternative to the current general DA-RT guidelines. CTSs would articulate guidelines for both authors and reviewers appropriate to diverse forms of (quantitative and qualitative) inquiry, guidelines that would be considered reasonable by the respective research communities (as well as promoting transparency consistent with the discipline's pluralism, according to the CTSs advocates). It seems to me that even if CTSs are apt for smaller communities, the general questions about transparency requirements (cf. Section 9.2.1) still apply. Next, are we not going to see discussions along the lines of Section 9.2.2 within the smaller communities too? Third, I wonder how generalist journals would deal with a variety of transparency conditions in the CTSs, how to decide which CTS is applicable? How to trade off quality between them? Will there be any implicit ranking or hierarchy between the CTSs?

   d   Besides these suggestions about qualifying the transparency idea in DA-RT, there are also critics of DA-RT who wonder why the focus should be so exclusively on "transparency and data access" or even whether we should have a transparency condition altogether. Why not highlight other standards like "intellectual engagement", which might also be a route to more transparency, or, "public relevance", what are the important and interesting questions the public wants to see addressed? (Those last two are a lot more social, social-epistemic, than DA-RT, the latter being more about rigorous reporting.) What is it in contemporary political science that selects research transparency and data access as the main focus? Even if it is a good thing to have transparency, the question according to those critics is whether it is really the most important problem or deserves the highest priority at the moment as well as whether an increased research transparency should be obtained via DA-RT. Why measure the quality of research with one decisive epistemic criterion like transparency?

In closing this section, let me highlight that I raised several critical questions concerning the condition of transparency. Even though it is an intuitively plausible condition that many philosophers support in dealing with values in science, specifying and implementing it in scientific practice presents a number of considerable challenges. The example of the DA-RT initiative in political science calls our attention to the negative impact transparency conditions might have on the social-epistemic constellation of a discipline and how it affects scientific plurality as I further discuss in the next section.

9.3 Social-Epistemic Practices and the Hampering of Scientific Pluralism: Ensuring Important Questions Can Be Addressed

Above, I mentioned the risk of *hampering scientific pluralism*. Let me explain what it means and why it would be bad for science. *Scientific pluralism* is a normative endorsement of there being *plurality* in science. Put differently, scientific pluralists defend that we do not only have de facto plurality in science (a descriptive claim), but that having plurality is a good thing (a normative claim). This plurality could be a plurality of forms of explanation, methods, kinds, styles of reasoning, systems of knowledge, and so on. Following scientific pluralism, it is important to make scientific practice, and the conditions it is supposed to live up to (e.g., Elliott's three conditions), congruent with the multiple goals science can have, the different interests and range(s) of questions it might have to address, as well as to avoid hampering or counteracting plurality in science.

In this chapter, I am not zeroing in on forms of explanation or kinds or styles of reasoning; rather I want to focus on social-epistemic constellations, analyze social-epistemic practices, and evaluate to what extent these practices hamper the consideration of certain questions, interests, goals, and methods. These social-epistemic practices could concern scientific journals and publishing in general, hierarchies in scientific disciplines, the establishment of disciplinary quality criteria, interactions between scientists and citizens, etc. The hampering could consist in directly blocking the entry for others that want to join, but also in sustaining a disciplinary structure that discourages interaction or promotes self-imposed isolation (cf. Van Bouwel 2009).

Why should we avoid *hampering* scientific pluralism? As I developed in earlier research, having a plurality of well-developed approaches available in a discipline helps to answer the different (kinds of) questions – and underlying interests/goals – scientists and other citizens (will) have in the best way possible – both accurately and adequately (Van Bouwel 2003, 2014). Moreover, in searching for the best possible questions and answers in science, having a plurality of approaches is necessary to ensure the mutual, transformative criticism desired (cf. Longino 2002). Thus, maximizing the number of questions addressable as well as the multiplicity of perspectives to ensure criticism seem imperative to get the best out of science.[6]

9.4 Scrutinizing Elliott's Second Condition: Representativeness

9.4.1 General Questions about the Representativeness Condition

Having sketched the general philosophically pluralist framework within which I scrutinize Elliott's conditions for justifiably bringing values into science, let us now consider Elliott's second condition, representativeness. The general idea is that values brought into science "should be *representative* of our major social and ethical priorities" (Elliott 2017, 10). In his book, Elliott offers us some examples of how to understand (the lack of) representativeness, for instance: (a) "the pharmaceutical industry's research priorities do not represent all the needs of the world's citizens well" (2017, 171); (b) manufactured doubts about tobacco use, industrial chemicals, and climate change represent concerns of a few wealthy corporations, not of the broader public (2017, 106); (c) Elliott writes in more general terms: "Industry values are likely to be less representative of the general public's interests ... when companies are testing the public-health or environmental impacts of products that generate a great deal of money for them" (ibid.). In these cases concerning public health and the environment, deciding what values should be considered (not) "representative" might be relatively easy. However, there might be cases in which it is more difficult to decide what values are "representative".

In order to tackle the more difficult cases, we first have to get a clearer view of *which* and/or *whose* values scientists or scientific communities should (justifiably) use and how exactly to understand Elliott's idea of representativeness? There is some ambiguity in Elliott's book.

A first interpretation of the idea of representativeness could be labeled as a *political* or *democratic* view focusing on *priorities*, advocating that the values used in the scientific process should be *democratically* justified. Elliott uses a vocabulary that links to democracy, i.e., a majority-idea, a general interest, considering what is representative of "our major social and ethical priorities" as opposed to special interests, not being representative. However, Elliott does not suggest any (democratic) procedure to decide what values are representative, or how opposing viewpoints about value judgments are to be accommodated, how scientists would be accountable, and other aspects that would be part of a well-developed political or democratic view. A second interpretation of the idea of representativeness could be labeled the *ethical* or "*depoliticized*" view with a focus on *principles*. For

Elliott, some values brought into science seem to be considered just not right. Is it because some values might not be in line with "our" ethical principles? This ethical approach does not focus on *priorities* but rather on *principles*, supposing some value judgments are just (in)correct. (Do they obtain their legitimacy by being well supported by reasons? How are reasons linked to representativeness?) In his 2018 précis of his book, Elliott writes:

> A second condition is that value influences should be *representative* of major ethical and social priorities. It is problematic when the values that influence scientific practice do not reflect important ethical principles (such as when research investments neglect the needs of those in low-income countries).
>
> (Elliott 2018, 4)

This quote shows the two interpretations, *priorities* and *principles*, combined.

These two views might be considered complementary. Elliott himself seems to follow an idea of complementarity and does not find the ambiguity *priorities/principles* problematic. He writes:

> When clear, widely recognized ethical principles are available, they should be used to guide the values that influence science. When ethical principles are less settled, science should be influenced as much as possible by values that represent broad societal priorities.
>
> (Elliott 2017, 14–15)

When can ethical principles be considered as "clear" or settled by scientists and how do they decide what the "broad societal priorities" are? Or, alternatively formulated, when can we conclude that value judgments are representative? Would this ambiguity not sometimes lead to conflicting guidance? What if *principles* override *priorities* or vice versa?[7]

9.4.2 The Representativeness Condition in the Social Sciences

Let us now further explore and modify Elliott's condition of representativeness in relation to scientific practice. As mentioned above, in his book he discusses the more straightforward examples concerning public health and the environment. However, sometimes scientific practice might be more complicated; there are many instances where the picture is more complex than having general versus special corporate interests. There might not be any agreement about "our major social and ethical priorities" or different interpretations about what our major priorities should be.

When looking at the social sciences, for instance, these different ideas about priorities, different values, interests, and goals, are reflected in there being different research approaches, "schools". When you address a certain topic, you might get many angles, different questions, and/or, at least, different answers.

Think, for instance, of experts addressing questions about the desirability of "free trade" (nicely tackled by Reiss 2019, in arguing against Jason Brennan's epistocracy, where Reiss shows how social scientists can have different priorities and values incarnated in different schools, and rationally disagree). In cases where you do not get different answers, where there seems to be a lot of agreement or talk about "consensus", at least be wary of social processes as I will now illustrate.

9.4.2.1 Representativeness, Consensus, and Hierarchy in Economics

Let us have a look at economics. What would representativeness mean in economics? Representing or choosing along our "major social and ethical priorities"? As listed in Section 9.1, we have to answer questions like: What research topics to prioritize? What specific questions to raise, methods to use, assumptions to make? What particular aims do we have in this specific context? How to deal with questions of uncertainty? How to report the research results, frame the conclusions? The questions and objects of investigation in economics are not objectively given, external, and unchanging, but rather context-dependent, constructed, and molded by values and disciplinary histories (see Rodrik 2015). However, notwithstanding the many choices that have to be made, there is still a surprising consensus in economics. Does that consensus mean that our major social and ethical priorities are being represented?

Sociological studies of economics teach us that the social-epistemic constellation of the field of economics is characterized by a pronounced internal hierarchy – most disciplines have some form of internal hierarchical structure but it is much more outspoken in economics (cf. Fourcade et al. 2015). Some of the aspects that stand out are (a) *the impact of the journal rankings*, especially the top five journals (cf. Heckman and Moktan 2018); (b) *the governance of the discipline is highly concentrated;*[8] (c) *concentrated publishing*: besides dominance in the governing bodies (and know, e.g., that the American Economic Association also controls several of the most important economics journals), the top departments make up a significant share of the authors and editors (also considering the social ties between them) in the high-prestige journals (cf. Wu 2007); (d) *agreement on quality criteria*: Fourcade, Ollion, and Algan (2015) also highlight that there is a strong internal agreement on quality criteria in economics, which results, for example, in a collectively organized recruitment process involving clear rank orderings. This is in contrast to other disciplines containing parallel hierarchies and rankings that interpret quality differently. There are no doubt more sociological characteristics we could highlight to illustrate the steep hierarchy within economics, but I hope the reader has gotten the point by now (let us also recall the danger linked to the transparency initiative in political science discussed above in instituting a journal hierarchy that suppresses certain methods and questions).[9]

What are the consequences of this social-epistemic constellation? James Heckman and Sidharth Moktan write about the tyranny of the top 5 (T5) journals in economics: "Using the T5 to screen the next generation of economists incentivizes professional incest and creates clientele effects whereby career-oriented authors appeal to the tastes of editors and biases of journals" (Heckman and Moktan 2018). The economist George Akerlof (who also won the Nobel Memorial Prize in Economic Sciences, just like Heckman) writes in relation to the top journals:

> The demands by the journals are just one more in a long series of previous demands for academic compliance that begin in high school, if not yet earlier. ... Of course, all that compliance, usefully, forces economists to master the field's current paradigm; those who wish to correct its omissions have special need for such understanding. But, just as there can be too little demand for compliance, there can also be so much that important problems are neglected: either because the problems themselves, or the best ways to tackle them, are deemed outside the frame of what is acceptable in the journals.
>
> (Akerlof 2020, 409–410)[10]

This social-epistemic constellation reinforces itself; the power concentration and the degree of asymmetry hamper the development of critical resources, mutual criticism, and new perspectives. It incentivizes young economists to follow the preferences and values of older dominant ones, limiting the extent to which research within economics considers a wide range of interests and values. It rather produces homogeneity, undermines broad scrutiny, and sustains an immature, misleading consensus.

So what would the value be of there being a consensus in economics?[11] In general, having a consensus among scientists might be a marker for the reliability of certain findings if reached independently by the respective scientists. However, when there seems to be a very small group of dominant economists trained at a very limited number of universities, one can question the independence of their opinions. So, does the consensus in economics signal that we have "our major priorities" represented? Or, alternatively, would a lack of consensus mean that the condition of representativeness is not fulfilled? Taking the social-epistemic characteristics in economics highlighted above into account, it might be doubted that consensus signals in this case that we have "our major priorities" represented. Moreover, might this social-epistemic constellation not be constraining certain kinds of questions from being raised and interests from being addressed?

One could consider particular social groups having questions and interests that are not well represented in the majority of the population. Their questions and interests may be excluded, marginalized, or otherwise not heard in some way – in such a way that they do not exist (becoming a case of *hermeneutical injustice* in Miranda Fricker's [2007] terminology). To steer clear of that scenario, we can evaluate social-epistemic practices (in line with Section 9.3) by looking at (a) whether they make it difficult for certain kinds of questions and interests to be considered – the questions and interests important to the particular social groups being underrepresented – as well as (b) whether they constrain (mutual) criticism of decisions about what questions and interests to consider.

9.4.2.2 An Example of Underrepresentation in Economics

For a clear example of certain kinds of questions and interests being constrained in economics, one can look at women being underrepresented in economics. Examining the social-epistemic constellation of the field of economics, one observes a high ratio of men to women in economics (women represent less than 15 percent of full professors in economics departments in the US, and around 20 percent in economics research).[12]

This striking gender imbalance, as well as female economists facing an unwelcoming environment (cf. Alice Wu 2017), (a) make it less likely that the concerns and questions of women are heard in economic research, meaning that the interests of an important segment of society often fail to be considered in determining in what direction economics has to be developed; and (b) limit the quality and scope of mutual criticism in the discipline as a unique and valuable perspective remains underdeveloped.

a That some questions only started to get serious attention once women were no longer absent in the discipline, has been well documented for the sciences as a whole as well as for economics in particular. For economics, think about the interest in typically female forms of labor, as Sharon Crasnow (2013, 414) writes:

> For example, in failing to include women when researching labor, domestic labor was not recognized as work. Seeing it as work alters the nature of the sociological study of human labor. Thus the feminist empiricism critique is more than a matter of "adding women and stirring".

More than applying the existing approach to new questions, the inclusion of women leads to a critique of the existing approach and the development of new perspectives. This brings us to (b).

b The absence of female scientists has caused the absence of particular forms of critical interaction, as historians of science have pointed out (see, e. g., Keller 1985). One might ask whether female economists add epistemically unique and valuable perspectives simply because they are women. We have already referred to the changes in the study of labor, but there have also been interesting studies about the differing perspectives of women and men economists. Ann Mari May, Mary McGarvey, and David Kucera (2018) found statistically significant differences in opinions (between male and female economists in Europe) in all five topic areas they examined: (1) market solutions versus government intervention; (2) environmental protection; (3) government spending, taxation, and redistribution; (4) core economic principles and methodology; (5) gender and equal opportunities. The largest gender differences involved areas (1) and (2). As for (1), the average female economist is less likely than the average male economist to prefer market solutions over government intervention, with the largest difference concerning views on whether stronger employment protection legislation results in weaker economic performance. Considering (2), women economists are more likely to support increased environmental protection than their male counterparts. I will not discuss further details of this study here; the main takeaway of this research is that the gender makeup of economics does affect outcomes and policy making. A greater representation of women would lead to a more diverse set of questions being asked and an alternative set of conclusions.[13]

A significant representation of dissenting perspectives is important for mutual criticism among perspectives, as Longino writes:

> A diversity of perspectives is necessary for vigorous and epistemically effective critical discourse. The social position or economic power of an individual or group in a community ought not determine who or what perspectives are taken seriously in that community. ... The point of the requirement is to ensure the exposure of hypotheses to the broadest range of criticism. ... Not only must potentially dissenting voices not be discounted; they must be cultivated.
>
> (Longino 2002, 131–132)

Increasing the number of perspectives that interact critically – exercise mutual criticism – raises the number of ways that knowledge can be challenged and tested. Considering the importance of critical interaction between well-developed alternative perspectives, as articulated by Longino, we must conclude that economics foregoes important opportunities of mutual criticism by leaving the female perspective underdeveloped and underutilized.[14] Incorporating more different, diverse perspectives increases the range and effectiveness of critical interaction in the discipline, positively affecting group decisions and deliberations, as empirical research has been backing up (see, e.g., Page 2007; Intemann 2009).

Thus, changing the distribution of perspectives (including values and power) within the discipline seems to be key to improving the collective epistemic outcome. Ensuring social epistemic practices that are open to a variety of different interests, questions, and goals is then to be understood as important for both the one being represented as well as for the one not being represented (with respect to a particular value) – there is a common interest to optimize mutual criticism.

9.4.2.3 Reconsidering Representativeness

Following this reasoning, I suggest a modification of Elliott's stipulation of representativeness. First, rather than "our major priorities" (and the ambiguity regarding *priorities* and *principles*, highlighted above), it seems important that the questions, values, and interests of diverse sets of the population are represented – some of which are clearly underrepresented in economics nowadays; the current consensus is detrimental. Second, in order to assure optimal mutual criticism, we should take into account the make-up of the discipline as a whole and ensure that a range of diverse approaches or perspectives is being sufficiently developed, not being hampered. Therefore, a focus on the distribution of perspectives is welcome.[15] Not ensuring optimal mutual criticism (or transformative criticism as Longino calls it) or not representing a range of important questions, implies an epistemic loss. Moreover, this mutual criticism could also affect what we understand as "our priorities": it could modify what we consider as most valuable, what turn out to be "false needs", or what values do not stand the test of being held accountable to available evidence.

What would it imply concretely for economics? Reconsidering representativeness along the lines I have sketched here supports a change from the steep internal hierarchy and (immature, misleading) consensus that hamper mutual criticism and hamper important questions being asked (and receiving the best possible answer) to a more level playing field, ensuring a social-epistemic constellation that is open to a variety of questions and epistemic goals. This resonates with demands to change the monistic economics curriculum for a more pluralist one put forward by student organizations like Rethinking Economics, Post-Crash Economics Society, Netzwerk Plurale Ökonomik, and PEPS-Economie. It is also in line with critiques brought forward in participatory exercises with citizens like the RSA's Citizens' Economic Council, highlighting the shortcomings of current economics.

These critiques and doubts uttered by the public as well as economics students vis-à-vis economics may be an instance of what Douglas describes here:

> Members of the public can dispute scientific claims because they think scientists are asking the wrong questions. ... social and ethical values legitimately shape the attention of scientists to certain topics or questions. But if what scientists care about asking does not align with what members of the public care about knowing, statements based on the findings can be greeted with skepticism, because the public thinks the scientists are not answering the crucial questions. ... what some citizens think are the crucial questions have not been well studied (as of yet). The values of those citizens and the values of the scientists are not aligning, producing skepticism about what scientists are reporting.
>
> (Douglas 2017, 91–92)

However, reconsidering representativeness in this way might raise worries. Would it not replace consensus with some form of isolationism among different approaches or extreme relativism? Would an alignment of values between certain scientific approaches and certain groups of citizens not eventually result in a number of isolated research approaches – each consisting of a particular scientific community teamed up with likeminded stakeholders – that are not engaging any longer with each other – no mutual criticism – as any scientific disagreement might be explained away as a difference in value judgment or in epistemic interest?

9.5 Social Epistemic Practices and the Hampering of Scientific Pluralism: (Epistemically) Productive Interaction between Approaches

In order to address these worries, we can explore what kind of social-epistemic interaction is needed to ensure representativeness without ending up in isolationism or extreme relativism. What norms or dynamics for scientific practice need to be lived up to in order to secure epistemically productive interaction between different approaches, representing a plurality of values?[16]

9.5.1 Democratic Models of Scientific Pluralism

In earlier work, I developed the idea that the interaction between different approaches can be made explicit in terms of models of democracy, focusing on three models in particular: consensual deliberative democracy, agonistic pluralism, and antagonism (cf. Van Bouwel 2009, 2015). Using these models of democracy helps us to clarify different interpretations of scientific pluralism by showing how they incarnate different ideals about the desired interaction – the social-epistemic practices – among the plurality of research approaches. It is a fruitful way to explore the social-epistemic constellations as well as highlight the close ties to democracy.

I will not discuss those three models in detail here, just briefly point at some key differences that might help in clarifying different social-epistemic norms in dealing with plurality in science. Let me first contrast the *consensual deliberative* model with the *agonistic* and *antagonistic* models of democracy:

Consensual Deliberative	Agonistic/Antagonistic
Consensus without exclusion	No consensus without exclusion
Diversity without dissent	Diversity and dissent
Eliminates conflict, depoliticization	Acknowledges power, keeps disagreement alive
Different views are complements	Different views can be substitutes

In her development of the idea of agonistic pluralism, Chantal Mouffe starts with two critiques of the consensual deliberative model of democracy. First, according to her the consensus cherished by the deliberative model of democracy is de facto not inclusive but oppressive. Following the Habermasian ideal of deliberation, it is necessary that the collective decision-making processes are so organized that the results are equally in the interest of all, representing an impartial standpoint. However, agonistic pluralists claim that the consensus-seeking decision making (in the common interest of all) conceals informal oppression under the guise of concern for all by disallowing dissent. Rawls and Habermas present their model of democracy as the one that would be chosen by every rational and moral individual in idealized conditions. In that sense there is no place for dissent or disagreement – the one that disagrees is irrational or immoral; the political has been eliminated, according to Mouffe (2005, 121–2).

Second, the theory of deliberative democracy eliminates conflict and fails to keep contestation alive. Agonistic pluralism points out that the consensual deliberative model can accommodate pluralism only by a strategy of depoliticization. We have to acknowledge power instead of ideally eliminating power and conflict, according to Mouffe:

> In a democratic polity, conflicts and confrontations, far from being signs of imperfection, are the guarantee that democracy is alive and inhabited by pluralism. ... This is why its survival depends on the possibility of forming collective political identities around clearly differentiated positions and the choice among real alternatives.
>
> (Mouffe 2000b, 4)

For Mouffe, political contestation should be a continuous practice, avoiding the oppressive consensus (seemingly in the interest of all) and acknowledging the plurality of values and interests (valued positively).

On the basis of these critiques of deliberative democracy, promoting a form of consensual pluralism, Mouffe develops agonistic pluralism. It emphasizes that the dimension of power is ineradicable in democracy, questions the ideal of consensus, puts contestation and antagonism central, and values pluralism positively: "the type of pluralism that I am advocating gives a positive status to differences and questions the objective of unanimity and homogeneity, which is always revealed as fictitious and based on acts of exclusion" (Mouffe 2000a, 19).

In Van Bouwel (2009), I used the agonistic model of democracy to make Helen Longino's norms for critical interaction among research approaches more explicit. Longino's account is often considered to be an example of a Habermasian, deliberative account, but I disagree and highlighted the agonistic elements in it.[17] The focus was on contrasting *consensual deliberative* with *agonistic pluralism*.

9.5.2 The Importance of Agonistic Channels

Now[18] we should probably look more at the contrast between *agonistic pluralism* and *antagonism*. The political climate being more polarized and antagonistic in the 2010s was itself a consequence of the consensualism of the 1990s, according to Mouffe. She suggests that rather than seeking an impossible consensus, a democracy should enable and institutionalize a plurality of antagonistic positions in order to create agonistic relations.

Agonistic Pluralism	Antagonism
Common symbolic space	No common symbolic space
Adversaries	Enemies
Interaction/confrontation	Incommensurability
Boundaries necessary but not reified	Reified boundaries

Agonistic relations differ from antagonistic relations to the extent that antagonism denies every possibility of a (stable or shaky) consensus between the plurality of positions; the "other" is seen as an enemy to be destroyed. Antagonists are not interested in finding common ground, but in conquering more ground, annexing or colonizing. Every party wants – in the end – to get rid of pluralism and install its own regime. Contrary to these antagonists, agonistic pluralists do cherish some form of common ground, a *conflictual consensus* in Mouffe's terminology, a *common symbolic space*, within which adherents of different research programs can engage, an arena where a *constructive channeling of conflicts* can take place. In this sense, agonists domesticate antagonism, so that opposing positions confront one another as *adversaries* who respect their right to differ, rather than *enemies* who seek to obliterate one another.

Thus, in order to domesticate antagonism, there is a need for a *common symbolic space* and for *agonistic channels* to express grievances and dissent. Lack of those tends to create the conditions for the emergence of antagonism, according to Mouffe. Agonistic pluralists aim at avoiding the antagonistic constellation, looking for ways to transform antagonism into agonism – providing a framework or agonistic channels that deal with conflicts through an agonistic confrontation. That could be a way to think of representativeness and the plurality of values. When understood in terms of social-epistemic constellations, it is the agonistic confrontation that brings about the desired (epistemically) productive interaction between a plurality of approaches, rather than consensual deliberation or antagonism.

Now, how could we understand the idea of agonistic channels in science? Let me suggest at least four arenas: (a) making explanation-seeking questions explicit, i.e., specifying research questions within a common symbolic space when debating what is the better explanation; you learn about the respective epistemic interests, the underlying values, and you make disagreement explicit as well as debate which values are legitimate (cf. Van Bouwel 2014); (b) citizen engagement with science (cf. Van Bouwel and Van Oudheusden 2017); (c) joint symposia; (d) scientific journals.

9.6 Scientific Pluralism and Agonism in Political Science

Let us finish by briefly looking at scientific journals as agonistic channels and return to the DA-RT discussion in political science. There exists an APSA journal, called *Perspectives on Politics*, that was started up in 2003 in response to Mr. Perestroika (just like the student organizations in economics mentioned above, Mr. Perestroika was a movement of researchers calling for more pluralism in political science). In Section 9.2.1, I mentioned the 2014 Journal Editors' Transparency Statement (JETS). The (then) editor of *Perspectives on Politics*, Jeffrey Isaac, did not sign JETS, considering it a threat to pluralism and writing that the journal "was the hard-won achievement of a coalition of academic political scientists seeking greater openness and pluralism within the discipline", an example of keeping the agonistic channels open.

Isaac is not arguing against transparency, but rather questioning whether a lack of transparency is really the main problem contemporary political science faces and whether imposing the new norms of transparency does not come at a substantial cost to intellectual vigor and the willingness to take intellectual risks in tackling interesting and important questions. Isaac writes:

> The one-size-fits-all expectations articulated in the DA-RT statement do not fit much of what *Perspectives on Politics* publishes. The strong prescriptivism of the statement is too heavy-handed to suit our journal's eclecticism. Perhaps most importantly, our journal operates on a distinctive epistemic premise: that the primary "problem" with political science research today is not that there needs to be more replicable research, "replication studies," and specialized inquiries, but that there needs to be broader discussions, and research projects, that center on ideas that are interesting and important.
>
> (Isaac 2015, 276)

Isaac advocates for a political science that is driven by what is interesting, what questions are important, rather than by applying a preferred method or strictly abiding by DA-RT prescriptions.

Isaac's plea resonates with my understanding of the representativeness condition as he emphasizes the importance of political science addressing "public concerns, public interests, and public groups to which political science speaks and ought to speak" (2015, 277) and he writes that what political science most needs now is "new and interesting work that speaks to the real political concerns facing the students we teach – and most of us spend most of our professional time teaching students – and the world in which we live" (ibid.). What research questions are neglected in current political science?

The emphasis on real political concerns and neglected research questions rather than on DA-RT does not imply that it necessarily comes at the cost of transparency. Transparency might be brought about through mutual criticism in a social-epistemic constellation like the one sketched above following Longino's work, rather than through a list of rules that according to Isaac produces professional narrowness and intellectual exclusion. The picture that arises is one of a question-driven scientific pluralism that is keeping the agonistic channels open to avoid ending up with a number of isolated, non-interacting research approaches, as well as to ensure interaction with adversaries to illuminate value-laden aspects of scientific practice and to consider how best to deal with them.

A last issue that remains for Isaac is how we will decide what the public concerns are, what are the important research questions? Isaac invokes the work of John Dewey to tackle this issue: "how should political scientists, and political science as an organized discipline, relate to – speak to, but also listen to – the complex and power-infused world that it both inhabits and takes as its object of study?" (Isaac 2015, 277). Dewey (1927) himself called for a continuing and lively dialogue between social scientists and the general public, saying the shoemaker knows how to make a shoe, but it is the public who knows where the shoe pinches. Such a continuing dialogue between social scientists and the public about what the questions should be, what convincing evidence is available to answer them, as well as why some values would be more important than others, would require a political science discipline that is not standing apart from society, but rather engaged with it, which brings us to Kevin Elliott's third condition, scrutinizing value influences through *engagement* between different stakeholders. Due to space constraints, we will have to leave the discussion of this third condition for another time.

9.7 Conclusion

With the help of debates in political science and economics, I highlighted the strengths and weaknesses of Elliott's first two conditions for justifiably bringing values into science: are these conditions unambiguous? How could they be implemented? How would they affect scientific practice? What do they teach us about the relations between science and democracy?

As I argued in this chapter, in order to answer different (kinds of) research questions – and underlying interests and values – of scientists and other citizens in the best way possible, we have to consider the social-epistemic constellation, look at the social-epistemic practices and asymmetries of power that might hamper or distort answering questions effectively. Making social-epistemic constellations in science better has to be understood as both a commitment to democracy – *in casu*, agonistic pluralism, limiting the inequality of voice – as well as to increased epistemic productivity – answering more questions in the best way possible – rather than the commitment to one particular perspective, implying epistemic loss. This requires that attention is being paid to the value judgments tacitly embedded in science, which I did in this chapter for political science and mainstream economics, with the aspiration to be value-neutral being a value judgment too.

To answer the question raised in the title, *are transparency and representativeness of values hampering scientific pluralism?*, I conclude that in some understandings *transparency* and *representativeness* do hamper scientific pluralism and prevent us from getting the best out of science as shown by the examples in this chapter. In response, I developed a modified understanding of *representativeness* with an emphasis on social-epistemic constellations that could also benefit *transparency* (without having to codify it in a set of strict rules to follow) while avoiding the hampering of scientific pluralism.

Notes

1 Besides the literature that is discussed below, other important work analyzing how values play a role in science include Brown (2013), Douglas (2009), Kincaid, Dupré, and Wylie (2007), and Zahle (2018).

2 Let me also add the definitions of *value* and *value judgements* Elliott (2017, 11) is using: "Broadly speaking, a value is something that is desirable or worthy of pursuit"; "value judgements are scientific choices that cannot be decided solely by appealing to evidence and logic" (2017, 12).

3 Besides the questions about incompatibility with epistemological diversity, DA-RT also raised questions about ethical, legal, and practical aspects or constraints under which many qualitative researchers have to work. One can wonder about how to achieve transparency while meeting ethical and legal obligations to protect human subjects, for instance, in cases where clandestine political opponents in an authoritarian regime inform one's research. In this chapter, I focus on the epistemological aspects, not the ethical or legal ones.

4 I question in the first place the transparency condition as understood by the DA-RT initiative here, a condition that is stipulated in an Ethics Guide. However, one could value transparency without seeing a need for DA-RT; transparency might perhaps be constituted or be brought about by social interaction or mutual criticism (cf. below). The latter might then be more epistemologically neutral, or, at least, less hampering.

5 I write *important* journals as there is a social-epistemic constellation to be taken into account in which there is a hierarchy between journals. I return to the impact of hierarchy when discussing representativeness in Section 9.4.

6 The philosophical literature in support of scientific pluralism has been growing fast over the last decade to the extent that scientific pluralism might be the predominant position in current philosophy of science; see, e.g., Chang (2012) and Ruphy (2017).

7 The ambiguity in the values in science debate is also discussed by Schroeder (2019) along comparable lines, opposing a *political philosophy approach* with an *ethics approach*.

8 Fourcade, Ollion, and Algan (2015, 12) find that in the period 2010–14, 72 percent of the American Economic Association's executive committee came from the

top five ranked departments and no committee members from outside the top 20 departments. Compare this to 12 percent for the American Political Science Association (APSA) and 20 percent for the American Sociological Association (ASA).

9 For a thorough analysis of hierarchy in economics and the social-epistemic consequences thereof, I refer the reader to Wright (2018). I only mention some of the central findings here.

10 This is an instance of the relation between the hierarchy/ranking of journals and the problems/questions addressed as discussed in Section 9.2.3.

11 Talking about consensus in economics often triggers unproductive debates – with mainstream economists pointing at disagreements among themselves and heterodox economists emphasizing the amount of agreement and lack of plurality in the mainstream. In order to avoid this kind of debate, I see at least two alternatives. One, compare economics with other disciplines (sociology, political science, etc.) as done by Fourcade, Ollion, and Algan (2015). Second, try to reason in less absolute terms, and rather in terms of *more* or *less* pluralistic. How to make the discipline more pluralistic, aiming to get rid of social-epistemic blockages, of what hampers productive interaction?

12 See www.aeaweb.org/about-aea/committees/cswep/survey for the statistics on professors. As concerns representation in research, Chari and Goldsmith-Pinkham (2017) "document the representation of female economists on the conference programs at the NBER Summer Institute from 2001–2016. Over the period from 2013–2016, women made up 20.6 percent of all authors on scheduled papers. However, there was large dispersion across programs, with the share of female authors ranging from 7.3 percent to 47.7 percent. While the average share of women rose slightly from 18.5 percent since 2001–2004, a persistent gap between finance, macroeconomics and microeconomics subfields remains, with women consisting of 14.4 percent of authors in finance, 16.3 percent of authors in macroeconomics, and 25.9 percent of authors in microeconomics".

13 May, McGarvey, and Whaples (2014) did a similar study in the United States also showing that the average male and female economists are reaching different conclusions on a number of central economic topics. Also see Mansfield, Mutz, and Silver (2015).

14 I elaborate the example of a female perspective here, but there are no doubt other particular interests that are being made invisible because of current, dominant social-epistemic practices. One could think, e.g., of a working-class perspective (a community more interested in securing employment opportunities to be explicitly targeted by economics than economic growth) or a sustainability perspective.

15 It does not mean that distribution of perspectives within the scientific community is the only issue to pay attention to; reliable methods for evidence gathering and analysis are obviously important too. The point is rather that the value of contributions of individual scientists depends also on the make-up of the community, therefore it is imperative to analyze how certain social-epistemic constellations hamper scientific pluralism.

16 *Epistemically productive* is being understood in terms of our capacity to answer our questions effectively, i.e., answering more questions in the best way possible as well as minimizing epistemic loss.

17 More recent contributions to the deliberative democracy literature have taken into account critiques like Mouffe's and are paying more attention to power differences, conflict, and dissent; cf. Curato, Hammond, and Min (2019).

18 "Now" both referring to the polarized state of politics dominant around the world in the year 2019 as well as now, at this stage of the chapter, considering the risk of isolation of approaches within a discipline.

References

Akerlof, G. 2020. "Sins of Omission and the Practice of Economics." *Journal of Economic Literature* 58 (2): 405–418.

APSA Committee on Professional Ethics, Rights and Freedoms. 2012. *A Guide to Professional Ethics in Political Science*. The American Political Science Association.

Brown, M. 2013. "Values in Science beyond Underdetermination and Inductive Risk." *Philosophy of Science* 80 (5): 829–839.

Chang, H. 2012. *Is Water H2O? Evidence, Realism and Pluralism*. Berlin: Springer.

Chari, A., and P. Goldsmith-Pinkham. 2017. *Gender Representation in Economics Across Topics and Time: Evidence from the NBER Summer Institute*. NBER Working Paper No. w23953.

Crasnow, S. 2013. "Feminist Philosophy of Science: Values and Objectivity." *Philosophy Compass* 8 (4): 413–423.

Curato, N., M. Hammond, and J.B. Min. 2019. *Power in Deliberative Democracy: Norms, Forums, Systems*. Basingstoke: Palgrave Macmillan.

Douglas, H. 2009. *Science, Policy, and the Value-Free Ideal*. Pittsburgh, PA: University of Pittsburgh Press.

Douglas, H. 2017. "Science, Values, and Citizens." In *Eppur Si Muove: Doing History and Philosophy of Science with Peter Machamer*, edited by M. Adams *et al.*, pp. 83–96. Dordrecht: Springer.

Elliott, K. 2017. *A Tapestry of Values: An Introduction to Values in Science*. Oxford: Oxford University Press.

Elliott, K. 2018. "Précis of *A Tapestry of Values: An Introduction to Values in Science*." *Philosophy, Theory, and Practice in Biology* 10 (7): 1–6.

Elliott, K., A. McCright, S. Allen, and T. Dietz. 2017. "Values in Environmental Research: Citizens' Views of Scientists who Acknowledge Values." *PLOS One* 12 (10): e0186049.

Elman, C., D. Kapiszewski, and A. Lupia. 2018. "Transparent Social Inquiry: Implications for Political Science." *Annual Review of Political Science* 21: 29–47.

Fourcade, M., E. Ollion, and Y. Algan. 2015. "The Superiority of Economists." *Journal of Economic Perspectives* 29 (1): 89–114.

Fricker, M. 2007. *Epistemic Injustice: Power and the Ethics of Knowing*. Oxford: Oxford University Press.

Heckman, J., and S. Moktan. 2018. "The Tyranny of the Top Five Journals." Available online at www.ineteconomics.org/perspectives/blog/the-tyranny-of-the-top-five-journals.

Hicks, D., P. Wouters, L. Waltman, S. de Rijcke, and I. Rafols. 2015. "Bibliometrics: The Leiden Manifesto for Research Metrics." *Nature* 520: 429–431.

Intemann, K. 2009. "Why Diversity Matters: Understanding and Applying the Diversity Component of the National Science Foundation's Broader Impacts Criterion." *Social Epistemology* 23 (3–4): 249–266.

Isaac, J. 2015. "For a More *Public* Political Science" *Perspectives on Politics* 13 (2): 269–283.

Keller, E. 1985. *Reflections on Gender and Science*. New Haven, CT: Yale University Press.

Kincaid, H., J. Dupré, and A. Wylie (eds.). 2007. *Value-Free Science? Ideals and Illusions*. Oxford: Oxford University Press.

Kreuzer, M., and C. Parsons. 2018. *Epistemological and Ontological Priors: Varieties of Explicitness and Research Integrity. Final Report of QTD Working Group I.1, Subgroup 1*. Available online at www.qualtd.net.

Longino, H. 2002. *The Fate of Knowledge*. Princeton, NJ: Princeton University Press.

Lupia, A., and C. Elman. 2014. "Openness in Political Science: Data Access and Research Transparency: Introduction." *PS: Political Science and Politics* 47 (1): 19–42.

Mansfield, E., D. Mutz, and L. Silver. 2015. "Men, Women, Trade, and Free Markets." *International Studies Quarterly* 59 (2): 303–315. May, A.M., M. McGarvey, and D. Kucera. 2018. "Gender and European Economic Policy: A Survey of the Views of European Economists on Contemporary Economic Policy." *Kyklos* 71 (1): 162–183.

May, A.M., M. McGarvey, and R. Whaples. 2014. "Are Disagreements among Male and Female Economists Marginal at Best? A Survey of AEA Members and their Views on Economics." *Contemporary Economic Policy* 32 (1): 111–132.

Monroe, K.R. 2018. "The Rush to Transparency: DA-RT and the Potential Dangers for Qualitative Research." *Perspectives on Politics* 16 (1): 141–148.

Mouffe, C. 2000a. *The Democratic Paradox*. London: Verso.

Mouffe, C. 2000b. "Politics and Passions: The Stakes of Democracy." Available online at www.politeia-conferentie.be/viewpic.php?LAN=N&TABLE=DOCS& ID=viewpic.php?LAN=N&TABLE=DOCS&ID=124. Retrieved August 12, 2019.

Mouffe, C. 2005. *On the Political*. London: Routledge.

Nagel, E. 1961. *The Structure of Science: Problems in the Logic of Scientific Explanation*. New York: Harcourt.

Page, S. 2007. *The Difference: How the Power of Diversity Creates Better Groups, Firms, Schools, and Societies*. Princeton, NJ: Princeton University Press.

Reiss, J. 2017. "Fact-Value Entanglement in Positive Economics." *Journal of Economic Methodology* 24 (2): 134–149.

Reiss, J. 2019. "Expertise, Agreement, and the Nature of Social Scientific Facts or: Against Epistocracy." *Social Epistemology* 33 (2): 183–192.

Rodrik, D. 2015. *Economics Rules: The Rights and Wrongs of the Dismal Science*. New York: W.W. Norton.

Ruphy, S. 2017. *Scientific Pluralism Reconsidered. A New Approach to the (Dis) Unity of Science*. Pittsburgh, PA: Pittsburgh University Press.

Schroeder, A. 2019. "*Values in Science: Ethical vs. Political Approaches.*" Paper presented at CLMPST2019, Prague.

Van Bouwel, J. 2003. "Verklaringspluralisme in de sociale wetenschappen." PhD thesis, Ghent University.

Van Bouwel, J. 2009. "The Problem with(out) Consensus. The Scientific Consensus, Deliberative Democracy and Agonistic Pluralism." In *The Social Sciences and Democracy*, edited by J. Van Bouwel, pp. 121–142. Basingstoke: Palgrave Macmillan.

Van Bouwel, J. 2014. "Explanatory Strategies beyond the Individualism/Holism Debate." In *Rethinking the Individualism/Holism Debate: Essays in the Philosophy of Social Science*, edited by J. Zahle and F. Collin, pp. 153–175. Dordrecht: Springer.

Van Bouwel, J. 2015. "Towards Democratic Models of Science. The Case of Scientific Pluralism." *Perspectives on Science* 23: 149–172.

Van Bouwel, J., and M. Van Oudheusden. 2017. "Participation With or Without Consensus? Technology Assessments, Consensus Conferences, and Democratic Modulation." *Social Epistemology* 31 (6): 497–513.

Wright, J. 2018. "Pluralism and Social Epistemology in Economics." PhD thesis, Cambridge University.

Wu, A. 2017. "Gender Stereotyping in Academia: Evidence from Economics Job Market Rumors Forum." Undergraduate thesis, UC Berkeley.

Wu, S. 2007. "Recent Publishing Trends at the AER, JPE and QJE." *Applied Economics Letters* 14 (1): 59–63.

Yamamoto, Y. 2012. "Values, Objectivity and Credibility of Scientists in a Contentious Natural Resource Debate." *Public Understanding of Science* 21 (1): 101–125.

Zahle, J. 2018. "Values and Data Collection in Social Research." *Philosophy of Science* 85: 144–163.

10 Max Weber's Value Judgment and the Problem of Science Policy Making

Lidia Godek

INSTITUTE OF PHILOSOPHY, ADAM MICKIEWICZ UNIVERSITY, POZNAŃ

> "Value" – that unfortunate child of misery of our science.
>
> Max Weber

10.1 Introduction

The keen interest in Max Weber's concept of value judgment focuses around both the evidence of its unique nature and the need for binding answers to the classic questions about the nature and meaning of axiological (i.e., normative) involvement in scientific practice. The pertinent question that requires an answer is: how should the role of value judgments be understood and to what extent are any value judgments acceptable (admissible) in scientific practice?[1] This is interesting not only on account of an appropriate interpretation of Weber's very thought, but precisely because of the clear and adequate self-awareness of social sciences in their own right. In my view, the status of values is undoubtedly one of the most important threads in Weber's reflection. It encapsulates the sense of the whole Weberian concept of science as vocation. But this is also why this issue raises particular difficulties in its interpretation that must be addressed to properly evaluate the concept of science and the scientific practice itself.

The question of undeniable strengths and possible weaknesses of Weber's concept of value judgments (especially the thesis of value neutrality, which is widely discussed to this day) may be considered on at least two levels: on the one hand, (a) the concept of methods for constructing scientific knowledge; and on the other (b) certain considerations on the very concept of science. As far as the former is concerned, Weber primarily focuses on analyzing the relationships between relevance to values, practical value judgments, and the postulate of value freedom. As for the latter, Weber considers the dimension of vocation and formulates the conditions to professionalize scientific practice.

In this chapter, I aim to present and examine Weber's views on the roles of value judgments on both methodological and practical levels. This dual perspective on value judgments can be seen both in his category of science as vocation (where it takes the form of internal and external conditions of science) and in the Weberian account of the institutionalization of science. I will examine the question of the interrelationship between theoretical and practical evaluation, which will allow me to subsequently demonstrate the coherence of Weber's reflection on science, in its methodological, cultural, and institutional aspects.[2]

First, in Section 10.2, I will outline the main methodological postulates in Weber's theory. I will focus on their three aspects: selectivity, transcendental presupposition, and autonomy. I will then provide an account of the practical value judgments. In light of these considerations, my proposition is that it is possible to distinguish a particular category of value judgments – the theoretical value judgments – in Weber's theory. In Section 10.3, I will consider the concept of science as vocation with particular focus on the normative criteria of scientific investigation. This will then allow me to identify the three dimensions of the Weberian vocation. In Section 10.4, in the context of science policy making, I will analyze the Weberian account of university as institution and point to the diverse forms of co-existence between politics and science as forms of institutional practice that Weber considered. Finally, in Section 10.5 I present my conclusions.

10.2 Levels of Evaluation

As is commonly known, the fundamental levels of the Weberian concept of science include considerations on the methods of scientific cognition and reflection on the importance of science in culture. These elements of Weberian thought have become a source of controversy, in particular with regard to the methodology of social science, value neutrality, and the problematic status of evaluation and value judgments.

The Weberian concept of evaluation arises on at least two different levels of thinking (Weber 1917/1949; Weber 1904/1949). On the first, methodological, level the meaning of evaluation is entirely subordinated to the tradition of doing social science. It is at this level that Weber places the selectivity of human cognition, both when it comes to selecting the subject of inquiry in terms of its cultural significance (the relevance to values, *Werbeziehung*), its selective description (the ideal type, *Idealtypus*), and its interpretation (the interpretive sociology, *Verstehende Soziologie*) (Weber 1917/1949, 21–22; 1904/1949, 81; 1922/1978, 1–10).

On the second level, Weber considers the process of practical assessment of phenomena. Weber confronts the scientific duties with value judgments founded upon ethical principles, cultural ideals, or a philosophical worldview. Here Weber firmly distinguishes between facts and values. It is also in this context that Weber considers the presupposition of modern science, that is, the postulate of value freedom (*Wertfreiheit*) as a source of scientific objectivity (Merz 1992, 185).[3] The postulate of value-free science, having a definite methodological sense, establishes a radical border between the scientific way of constructing the objects of cultural reality and practical evaluations. In other words, on the first level, the basic issues around the concept of evaluation relate to the method of knowing the cultural reality (Weber 1904/1949, 89–90). On the second level, on the other hand, the discussion focuses on the prohibition of integrating unconscious axiological (ideological) threads into the scientific process.

What most prominent authors underline is the insufficiency and the problematic nature of Weber's propositions aimed at explaining and justifying the different levels of evaluation (Nusser 1986; Jacobsen 1999; Bruun 2016), for there is a constant tension between the methodological dimension of evaluation and the evaluation from the worldview perspective, or between the relevance to values and value freedom, between "good" and "bad" evaluation. Most of the debate around value judgments brings to the fore their problematic cognitive status, and points to the challenges around their objectivity and empirical verification.[4]

In what follows, I will attempt to reconstruct in more detail the two levels of axiological reflection in Weber's theory.

1 **Methodology level.** Through the concept of relevance to values, Weber defines the path for the philosophical reflection on science. The "relevance to values" is the basic rule of philosophical interpretation of scientific interest, which determines the selection and formulation of the subject matter of empirical inquiry in terms of its cultural significance. This level addresses the following aspects:

a Selectivity. The particular nature of science rests in its selectivity. Values play the role of a criterion for selecting out what is significant from the viewpoint of a given cultural formation for its scientific account.

> All the analysis of infinite reality which the finite human mind can conduct rests on the tacit assumption that only a finite portion of this reality constitutes the object of scientific investigation, and that only it is "important" in the sense of being "worthy of being known".
>
> (Weber 1904/1949, 72)

b Transcendental presupposition. Weber points to the normative presupposition of any cognitive activity that, like every product of human activity, is affected by the culture. Values, as a transcendental presupposition of human cognitive abilities, are relational: they "exist" between an individual (the investigator) and the object (a portion of the reality being examined):

> The transcendental presupposition of every cultural science lies not in our finding a certain culture or any "culture" in general to be valuable but rather in the fact that we are cultural beings endowed with the capacity and the will to take a deliberate attitude towards the world and to lend it significance. Whatever this significance may be, it will lead us to judge certain phenomena of human existence in its light and to respond to them as being (positively or negatively) meaningful.
>
> (Weber 1904/1949, 81)

Science constitutes the part of human activities that form our culture. A special task is therefore placed before philosophy to answer the fundamental question on the relation between values and empirical reality.

c Autonomy. The category of relevance to values establishes the autonomy of science since its purpose rests in realizing the truth as a fundamental cognitive value. The world of science is a domain of theoretical value judgments. Although Weber assumes that scientific objectivity must be associated with the subjective prerequisite of empirical knowledge, he also believes that recognizing the value of truth forms the basis of empirical knowledge in general:

> The objective validity of all empirical knowledge rests, and rests exclusively, upon the fact that the given reality is ordered according to categories which ... are based on the presupposition of the value of that truth which empirical knowledge alone is able to give us. With the means available to our science, we have nothing to offer a person to whom this truth is of no value – and belief in the value of scientific truth is the product of certain cultures, and is not given to us by nature. He will, however, search in vain for another truth to take the place of science with respect to those features which it alone can provide: concepts and judgments that allow ... [empirical] reality to be ordered intellectually in a valid manner.
>
> (Weber 1904/1949, 110–111)

Truth, as scientific value, is a source of value judgments, which are neither empirical reality nor its representation. Rather, they facilitate the ordering of (Weber 1904/1949, 110), and give sense to, reality. Valid judgments are always based on the logical analysis of our perceptions (Weber 1904/1949, 107) by means of concepts (ideal types). Thus, agreeing with Kant, Weber questions the view that knowing depends solely on perceiving. Judgments play a crucial role in everything that defines the cognitive condition of human beings (Weber 1904/1949, 110–111).

2 **Worldview (practical) level.** Values are a platform for practical evaluations. Values carry a validation potential as they embed human activities and choices into axiological structures. The binding and benchmarking values serve as a criterion for evaluation. In this context, Weber mentions the "value axioms", which determine their validity. Activities constitute a kind of a platform that brings together the world of culture (as a domain of practical activities) and the realm of values. Although on several occasions Weber strongly emphasizes that a true philosophy of values is not able to capture the most important moment of a state of affairs, yet almost every viewpoint of reality and real people bears a texture of various spheres of values (Weber 1917/1949, 17). The world of life, as a domain of practical value judgments that follow from ethical premises, ideals of culture, or a specific view of the world (worldview), does not lend itself to any scientific

settlement but depends entirely on the evaluating attitude toward our own and other people's actions (Weber 1917/1949, 1–2).

The above account of evaluation is an undeniably positive contribution by Weber to axiological issues. Values provide the original criterion necessary for recognizing the transcendental conditions of knowing the culture. Weber's considerations reveal the distinct tension between the methodological and practical (worldview) dimensions inherent to all human thinking and bring to the fore the very problematic relations between the process of "value freedom" (axiological neutrality) and establishing science as a part of culture.

10.2.1 *Values as the Method and Subject of Scientific Inquiry*

What does it mean to know cultural reality? The Weberian concept of science does not only focus on the study of cultural reality but also, and perhaps more importantly, on the ways of perceiving and articulating it. At first glance, the systematic reconstruction of Weber's methodological background appears to demonstrate the complexity of his scientific instruments at every, even most abstract, level. The first category is "relevance to values", which is a key category seen by Weber as the building block behind the philosophical concept of science. By referring to this Rickertian category, Weber (1917/1949, 219) believes that cultural reality is based on values, while the method of constructing concepts, expressed in the form of ideal types, allows giving it a theoretical form.[5] In the context of methodology, this category demonstrates the particular nature of the scientific dimension of scientific knowledge (Burger 1976, 7).

On the methodological level, values play the role of a criterion for selecting out what is significant from the viewpoint of a given cultural formation in order to formulate its scientific description. The relevance to values relies on relating the object of scientific investigation to the cultural values, which define what is being investigated (Weber 1904/1949, 70). The specific cultural events emerge in the context of value ideas that give meaning to empirical reality. The concept of culture is a concept of value that spans only those parts of reality that become significant to us through this very connection that endows the facts with their "cultural significance" (*Kulturbedeutung*) (Weber 1904/1949, 81; 1930/2005, 13–14). The culture, metaphorically speaking, "is a finite segment of the meaningless infinity of the world process, a segment on which human beings confer meaning and significance". It should also be emphasized that it is impossible to determine what is significant to us by way of any unconditional empirical investigation. To the contrary, identifying what is significant is a condition for anything to become the subject of our investigation (Turner 2019, 582).

The core of the Weberian method consists in the process of constructing concepts to gain knowledge about cultural reality. It defines the way in which a researcher constructs and applies any concepts. The initial premise assumes that the particular nature of the cognitive purpose of any work in the field of social sciences consists in making it dependent on the "one-sidedness" of viewpoints that affect the choice of the object of investigation as well as its analysis and presentation (Weber 1904/1949, 72, 91).

Knowing the cultural reality always involves knowing from a "particular point of view", considered here as an elementary presupposition of any scientific investigation:[6]

> All knowledge of cultural reality, as may be seen, is always knowledge from particular points of view. When we require from the historian and social research worker as an elementary presupposition that they distinguish the important from the trivial and that he should have the necessary "point of view" for this distinction, we mean that they must understand how to relate the events of the real world consciously or unconsciously to universal "cultural values" and to select out those relationships which are significant for us.
>
> (Weber 1904/1949, 81–82)

The construction that allows capturing the essential features of any studied phenomenon is the Weberian ideal type, which should be included in the repertoire of any scientific inquiry. There are various "one-sided" viewpoints from which the phenomena important to us could be considered, so diverse principles can be applied when selecting out the dependencies upon which an ideal type is to be built. As Weber writes:

> For those phenomena which interest us as cultural phenomena are interesting to us with respect to very different kinds of evaluative ideas to which we relate them. Inasmuch as the "points of view" from which they can become significant for us are very diverse, the most varied criteria can be applied to the selection of the traits which are to enter into the construction of an ideal-typical view of a particular culture.
>
> (Weber 1904/1949, 91)

The Weberian concept of science excludes a single viewpoint, which means that its definition undergoes continuous change (science as an open project). On the one hand, this pluralism and freedom in the selection of possible viewpoints for the construction of ideal types is a consequence of the development of science subordinated to formal rationality. Formal rationality is defined here as a common agreement among researchers as to the validity of methodological forms being a prerequisite to any scientific work. On the other hand, pluralism and freedom transform into rules of doing science as an open project, albeit subordinated to the fundamental cognitive value of truth. Thus, the objectivity of science is determined by the inter-subjectively acceptable conditions of identifying methods of access to reality, and its axiological dimension by the supremacy of truth and intellectual freedom.

Furthermore, Weber emphasizes the cultural nature of scientific investigation, which implies a close relationship between a scientific inquiry and its cultural context. In cultural sciences, no one deals with issues that bear no significance for them (Weber 1904/1949, 72). However, the meaning of what is being investigated is not obvious and requires a scientific intervention. What matters is not the facts themselves but rather their meaning expressed by concepts. The only measure of success for an ideal type is, therefore, its instrumental handiness.[7] In addition, no knowledge of the cultural process is conceivable except for the knowledge that is based on the significance attached to the reality we live in (Weber 1904/1949, 80–81).[8]

So far, the above analysis has focused on the methodological role of values but, according to Weber's account, values can themselves become the subject matter of scientific inquiry. In other words, both the subject matter and the level of inquiry cannot be entirely free from any and all values. The process used to interpret values is referred to by Weber as the discussion of value judgments (Weber, 1917/1949, 21–22). According to Weber, it consists in analyzing individual activities and discovering the principally evaluative ways of perception that are revealed through those activities (Bruun and Whimster 2012, xxii). It aims at identifying the consequences that follow from those value axioms that are accepted by human beings as assumptions (motivations) underlying their actions. The sense of the discussion of value judgments is reduced by Weber to four main elements (Weber 1917/1949, 20–21):

1 *Identifying value axioms.* The process Weber has in mind here is an operation that attempts to identify those values that form the foundation of value judgments based on the analysis of their sense. The result of this operation is not to discover any facts but to recognize the underlying values.

2 *Deducing the potential implications* that follow from the accepted value axioms when they form the foundation of practical value judgments. This means that the process of analyzing any practical evaluation seeks to identify the implications of practical value judgments.

3 *Deducing the factual implications*, which follows from adopting a practical value relationship by associating those value judgments with the specific and indispensable means, and identifying those side-effects that are not necessarily expected.

4 *Identifying new value axioms.* The discussions of practical value judgments can reveal new value axioms that have not been taken into account by the proponents of practical evaluation.

In summary, the Weberian vision of science is characterized by the following properties:

1 Science should not strive to establish any consolidating, unifying investigation perspective; instead, it should accept the inevitable cognitive perspectivism ("one-sided" points of view).

2 Science should not deal with the things in themselves but focus instead on the conceptual constructs and their interconnections, which are subjected

to a constant "problematization" (it is all about connecting problems and not about things in themselves):

> It is not the "actual" interconnections of "things" but the conceptual interconnections of problems that define the scope of the various sciences. A new "science" emerges where new problems are pursued by new methods and truths are thus discovered which open up significant new points of view.
>
> (Weber 1904/1949, 68)

3 Science should not deal with establishing any raw facts but their "meanings" that result from that problematization:

> There are ... "subject matter specialists" and "interpretative specialists". The fact-greedy gullet of the former can be filled only with legal documents, statistical work sheets and questionnaires, but he is insensitive to the refinement of a new idea. The gourmandise of the latter dulls his taste for facts by ever new intellectual subtilities.
>
> (Weber 1904/1949, 112; cf. Burger 1976, 76–77)

According to Weber, these three elementary properties of each investigation process – perspectivism,[9] conceptual constructivism, problematization – define the methodological dimension of science. We must analyze the concepts we apply (the ideal type), we must be aware of the circumstances in which we think (the culture), and, finally, we must know how the outcomes of our thinking affect our lives.

10.2.2 Practical Value Judgments

I will begin my analysis of the second level – the worldview values – by exploring the definition of practical value judgment. In principle, Weber presented the definition of practical value judgment in two places in "The Meaning of 'Ethical Neutrality' in Sociology and Economics". First, Weber states as follows: "*value judgments* are to be understood, where nothing else is implied or expressly stated, practical evaluations of the unsatisfactory or satisfactory character of phenomena subject to our influence" (Weber 1917/1949, 1). Second, he mentions "*practical* evaluations regarding the desirability or undesirability of social facts from ethical, cultural or other points of view" (Weber 1917/1949, 10). In emphasizing the practical elements (as opposed to those theoretical components), Weber characterizes, at first glance, the sphere of evaluation as opposed to science and thus confirms the logical gap between them (Bruun 2016, 67–68). For Weber, the sphere of evaluation is the realm of human practice. It seems that for this very reason he associates value judgments with the adjective *practical*. The practical value judgments escape the cognitive criteria of truth and falsity as well as any subsumption pursuant to the principles that underpin cognitive and logical judgments. Yet Weber recognizes the key role of evaluation in everything that is important to human beings in the sphere of practice (Weber 1917/1949, 1–3).

In general, any judgment involves the ability to think about what is particularly based on a universal rule (cultural or ethical). Evaluating relates, therefore, to the ability of reasonably subsuming a particular detail (a fact) under a rule or to determine whether something falls under a given rule or not. Each value judgment is therefore brought about through recognizing a particular value as a rule. The value judgments that Weber opposes are those judgments that subsume the subject of our scientific inquiry under the rules that originate in ethics, politics, religion, or economy. Although, in a logical sense, Weber eliminates evaluation from the domain of science, for purposes of practical goals or objectives of science itself, evaluation remains necessary. This opens the doors to evaluation in science, a possibility which Weber does not at all exclude. It is rather a question of the extent to which evaluation can be allowed. If, through scientific inquiry (whether through empirical analysis or understanding explanation), we identify certain rules pursuant to which an individual undertakes a specific activity, we cannot infer from those rules any specific cultural content as the only basis of evaluation. Science does not assign the quality or power of absolute ethical rules to any cultural values. In the discussion of values (interpretation of values), different value judgments are being confronted. Any valid evaluation of human acts can be criticized on the basis of a different worldview. The value axioms are at the heart of any practical act of will. In other words, scientific analysis consists in identifying those value axioms and demonstrating the relationships that exist between them:

> Philosophical disciplines can go further and lay bare the "meaning" of evaluations, i.e., their ultimate meaningful structure and their meaningful consequences, in other words, they can indicate their "place" within the totality of all the possible "ultimate" evaluations and delimit their spheres of meaningful validity.
>
> (Weber 1917/1949, 18)

The actual objection by Weber, therefore, does not concern values in themselves but evaluation on the basis of a worldview, which assigns a legitimizing function to values. Similarly, the aim of Weber's methodological thought is not to eliminate all values from science, but to uncover those axiological determinants of science that are culturally conditioned. However, for Weber the axiological presuppositions of science (as a cultural product) and evaluating the results of a scientific inquiry are two entirely separate matters. On the one hand, he analyzes the relationship between practical value judgments and the relevance to values, which describes the two dimensions in which values are present in science. On the other hand,

Weber investigates the relationship between practical value judgments and the interpretation of values, which establishes the autonomy of any scientific analysis of values (Bruun 2016, chapter 3).

10.3 Three Dimensions of Science as Vocation

As demonstrated above, the question of ideological values in science is not limited to theoretical or methodological issues but acquires an important ethical and social-political dimension. Weber's philosophical criticism of the lack of honesty in presenting scientific results allies, in a particular way, with a certain ethos of thinking, which aims to free science from the power of political, ideological, or economic powers. Weber's postulate of scientific rationality is consistently complemented by his program of science as vocation. Not only does Weber notice the culture-forming and civilizational role of science but, crucially, he realizes the potentially destructive influence of economy, religion, or politics (which are often "inspired" by ideology) on science. The ideological values in science can become a source and means of political and economic domination. While the enslavement of science by ideology, politics, economy etc. is relatively evident (the pressure is external as it comes from outside science), the situation is entirely different when it comes to the pressure from within science. This is when science can be deprived of control "sensors" or controls shaped by the pattern of "vocation".

When giving his account of modern science, Weber points to two types of conditions: (1) the external conditions that he associates with the process of "disenchanting the world" and rationalization (the professionalization and specialization in science); and (2) the internal conditions among which he places the category of vocation. At the very center of this polarization lies the question about the relationship between the sphere of values and practice (the instrumentalization of values). The rationalization process has led not only to the different ways of dealing with reality (the methodology of scientific life) but also to the fragmenting of the realm of values (the values of science as opposed to the values of economy, politics, aesthetics, and erotics) (Weber 1918–19/1946, 148).

In the context of methodology of scientific life and of the modern ethos of science formulated on its basis, it is important to be aware of the role of vocation in science and its components. First, the Weberian concept of vocation emphasizes the internal relationship between the two types of evaluation: substantive and axiological evaluation. Vocation associates the duty of cultivating knowledge in a substantively appropriate and methodologically correct manner with the values of scientific practice. Substantive and axiological evaluations are indeed co-necessary elements of human culture and the co-necessary parameters of human thinking at the level of cognitive operations and choices of human cognitive activities. Participating in science as vocation requires specific personality attributes that include a sense of responsibility, passion in the quest for truth (dedication to the cause) (Weber 1917/1949, 5; 1918–19/1946, 135), intellectual integrity in the application of methods of its verification (Weber 1917/1949, 3), and the discipline of reasoning (self-control) (Weber 1917/1949, 5). These axiological premises form the corpus of the scientist profession that can be defined as its ethos.

Second, the category of vocation refers to the supremacy of the standards of professional conduct over any benefits or self-interest. In this context, Weber mentions the requirement of independence of science and the freedom from interference from politics (Weber 1917/1949, 2–3).

But there is also a third, equally important, component to vocation. It is the culture that forms the strongest and deepest foundation of scientific intuition and creative imagination. Weber strongly rejects the presupposition that moral or political views can be used to achieve scientific goals (Weber 1918–19/1946, 152). He is also skeptical of the belief that science can be used to achieve moral or political goals. Yet he never directly claims that the effects or products of science are neutral toward the social world. It seems to him that scientific involvement in the world can contribute to making it better. Still, it is doubtful, according to Weber, whether any social involvement of science can yield better results for science itself. Intellectual honesty requires that no arbitrary meaning is attributed to science, nor should we endow it with the power of final settlement in practical matters. Science cannot help in resolving moral or political disputes in terms of legitimizing individual point of views. Rather, it can help to reconstruct our value systems (Weber 1917/1949, 19).

10.4 Models of Science Policy Making

The notion of university policy belongs to the key concepts in Weber's instrumentation. It is so not only because Weber considers the institutional dimensions of modernity (the modernization theory) but also – and perhaps most importantly – because it completes his vision of science. It implies that *science policy making* is not an attribute or aspect of modern science but is identical with it. If it is at all possible to give an account of science, it is only of science that has been institutionally shaped. When explaining the characteristic features of the Occident's culture, Weber focuses his attention on the institutionalized dimensions of change in various social systems, including science (Habermas 1982, 251). The modernization changes consisted in the occurrence of rational thinking and institutionalization of activities (Turner 1992, 11–12). On the social level, Weber analyzes various types of rationalization that simultaneously produce a form of the modern subjectivity of human beings as free, purposeful, and rational individuals. In addition, and inseparably to this, the rationalization creates a system of institutional safeguards of these freedoms. The institutionalization of science, and the forms it develops to master reality, become an indispensable portion of a wider whole that is the social process of modernization. Along with his investigation into the methodology of cultural sciences, Weber considered the question of the very structure of modern science (Lassman and Velody 2015, xiv). This resulted in his proposition that modern science has become institutionalized.

On the one hand, science appears to be a component of culture (next to technology, art, literature, law, morality, economy, or eroticism) that has developed from the very same traditional and metaphysical image of the world (Weber 1918–19/1946, 155). The main mechanism involves the autonomy of values, for science has its own standards in this regard, which it aims to defend against any ideology ("own standards of value"; Weber 1904/1949, 58). However, a rather complex situation follows as a result. Freeing science from ideological values secures the minimum conditions for its impartiality. Yet the very values of truth, reliability, soundness of argument, etc., safeguard the autonomy. So, when demanding autonomy for scientific values, Weber does not propose a strict separation between ethics and science (as indicated by Nusser 1986, 80). Although ethical values may conflict with the values of science, acting in the interest of values of science may demonstrate dignity, as well as determine their sense.

On the other hand, Weber sets out the rational techniques of realizing the values of science (the process of intellectualization) that require a certain institutional framework. The organized form of practicing science is the university. He also identifies the forms of modern subjectivity as autonomous and able to consciously respect values as the basis for justifying scientific inquiry and axiological orientation (Weber 1918–19/1946, 138–139).

An indication of such a close relationship between science and institutionalization implies a new understanding of the science policy that Weber describes by contrasting the two forms of the ethos of science that directly refer to the discussion of values. On the one hand, this university ethos rests in shaping human beings by instilling in them political, ethical, or aesthetic attitudes. On the other hand, it is founded on the specialized education underpinned by the virtue of intellectual integrity (Weber 1917/1949, 2–3).

Weber's broad interest in politics concerns the presence of politics in scientific practice and the question of the nature of this presence (Weber 1919/1946. 77). Weber's papers cover at least three perspectives of his account on the relationship between science and politics. Science can be defined as (i) the experience and symbolic communication free from any political interference; (ii) the capital (teleological approach) where the role of the state is reduced to only supporting those practices that are politically valuable and are in the public interest; and (iii) the political discourse where the state treats science as a means for promoting a particular ideology.

The first of the proposed approaches matches the category of vocation and, in this sense, constitutes an important ideal for Weber. Science as vocation is founded on certain internal beliefs and principles professed by the people institutionally involved in science. On this account, science is not a purely abstract or teleological concept, but an anthropological concept that describes a particular type of practice that is typical of the scientific profession. Its meaning is well illustrated by Weber's description of vocation with the three dimensions of its presence: ethics, self-control, and autonomy. It is only through the combination of these aspects of vocation that the relationship between the substantive and normative criteria of science can be fully grasped. The concept of science is based on the presupposition of convergence between the practices that are based on a common repertoire of principles, styles, values, and symbols, which make up the general universe of science. However, if the concept of science, as a common experience (also a historical experience), is complemented by imagining and symbolic representation, then it acquires an institutional meaning. Science as a form of social life is shaped by the networks of representation: scientific papers, certain behavioural codes, and structures (the university) that are organized by these representations.

However, the institutionalization of science has its drawbacks. Weber criticizes the scientist profession for (i) excessive specialization; (ii) a kind of "flocking"; (iii) scientific thinking in terms of resources required for solving specific problems (the technicization and instrumentalization); (iv) heteronomy as dependence on other social practices (politics, economics); and (v) goals that are external to science (goals that are subordinated to external principles of creation, exchanging thoughts, etc.). Weber also declares that

> [he] will struggle relentlessly against the severe self-deception which asserts that through the synthesis of several party points of view, or by following a line between them, practical norms of scientific validity can be arrived at. It is necessary to do this because, since this piece of self-deception tries to mask its own standards of value in relativistic terms, it is more dangerous to the freedom of research than the former naive faith of parties in the scientific "demonstrability" of their dogmas.
>
> (Weber 1904/1949, 58)

Vocation, as an internal condition that determines the nature of modern science, can be supplemented by certain external conditions stipulated by Weber (Weber 1918–19/1946, 129). One of them is the autonomy of the university and the guarantee provided by political institutions for the freedom of scientific inquiry. This autonomy, together with the freedom to announce the results of inquiry and the freedom to teach, constitute the freedom of science. The state is to act merely as a regulator (*the regulative science policy making*). It is there to provide the rules and legal guarantees for conducting scientific investigation. This is the essence of the third dimension of Weber's concept of freedom (next to the previously discussed methodological and practical levels). The historical development and the universalization of freedom turn out to coincide with the institutionalization in which the relationship between the institutional conditions of science and its integrity fully arises. The reason for this lies in the fact that the freedom of science and political freedom depend on one another (Gostmann 2015, 110). Weber did not seem to have any major doubts as to the political conditions under which science can be practiced autonomously and freely. In his opinion, despite many objections to modern democracy, the democratic system is the only one based on freedom and individualism. Democracy as a sophisticated form of freedom is compatible with "free reason" (Weber 1906, 347–348).

On the second level, science is defined in terms of its purpose. The social definition of science – as Weber understands it – refers to the meanings and values conveyed by science as well as by the institutions that represent it (Shapin 2019, 12–14). These values are closely linked with scientific practice. The fundamental objective of science is to provide information about our cognitive condition and the consequences of our actions. It follows that the teleological dimension of science cannot be reduced to deciding about the social value of science nor to commercializing its results (Weber 1918–19/1946, 131).

Thus, Weber opposes defining science as the core of political and economic development and treating it as a means of solving concrete problems. He would rather be inclined to define science as cultural capital realized through its critical potential:

Science today is a "vocation" organized in special disciplines in the service of self-clarification and knowledge of interrelated facts. It is not the gift of grace of seers and prophets dispensing sacred values and revelations, nor does it partake of the contemplation of sages and philosophers about the meaning of the universe. This, to be sure, is the inescapable condition of our historical situation. We cannot evade it so long as we remain true to ourselves.

(Weber 1918–19/1946, 152)

Science can, therefore, be a legitimate field of investment by states (*the protective science policy making*). In this scenario, state policy boils down to identifying those intellectual forms that deserve the most support as they constitute a means to achieving scientifically important goals.

Lastly, science can be understood as a tool in the service of political ideals (Weber 1918–19/1946, 152). This approach draws attention to the discursive construction of science as a subject of politics. It assumes the possibility of using science in achieving other goals (political goals), such as promoting *raison d'état*, national interest, social integration, economic development, etc. Weber rejects the presupposition that science can develop practical value judgments by deriving them from some "political worldview", or that it should provide justification to any such judgments. The fact that social sciences developed as an answer to practical impulses (the need to solve specific problems) tempted the connection between the purposes of science and specific political ventures: "the university is a state institution for the training of 'loyal' administrators" (Weber 1917/1949, 7). Weber rejects the above vision of science to the extent that any practical value judgments were to be used to set binding rules and ideals to formulate recipes for practical activity.

Incorporating practical value judgments into science endows it with the status of an "ethical science". On this account, the universities and other research institutions carry out a mission of special importance for the state and the nation and co-shape moral standards in public life (*the integrating science policy making*).

10.5 Concluding Remarks

The first objective of this chapter has been to identify the role of value judgments in science and to consider their admissibility in scientific inquiry and science policy making. My analysis has allowed me to identify the two levels of evaluation, which draw on methodological and practical (world-view) perspectives. On the first level, truth constitutes the basis of theoretical value judgments, while the worldview perspective forms the source of practical value judgments.

My goal has also been to shed more light on certain relationships between these methodological and practical dimensions. Two important conclusions have arisen in this regard. First, although a clear-cut boundary between scientific validity and evaluation is much desired, it is not fully achievable in practice. This is apparent in Weber's considerations on the methodological postulates of science and his identification of prerequisites for scientific validity.

Second, Weber formulates the concept of science as vocation and identifies the substantive and axiological prerequisites of science, which in itself demonstrates a false dichotomy between science deprived of any values (value-free science) and its entirely ideologized version. Although the idea of ideologized science is utterly implausible, it does not follow that science deprived of any value judgments is in fact its best possible alternative.

As is demonstrated by the category of vocation, science is able to reflect back on itself, on its own consequences, and is aware of the conditions under which doing science is at all possible. Science performs a critical function when, by questioning the forms of thinking outside science, it simultaneously reviews its own methods and axiological involvement, and modifies itself, being the subject of criticism. It is no longer just the scientific cognition but, at least, the relationship between the cognition and practical cultural action that, for the first time, becomes the main subject of philosophical thinking.

As it has been demonstrated in Section 10.4, despite the main purpose of Weber's argument aimed at defending the autonomy of science, he accepted, in fact, the various forms of coexistence of politics and science as forms of institutional practice. Weber did not aim to separate science from politics but to design such institutions that would safeguard the autonomy of science and its values. In my interpretation, Weber supported the idea of regulative science policy making which guarantees the freedom of scientific inquiry and freedom to teach.

The theoretical and philosophical paradigm developed by Weber has not lost its prominence or its cognitive value His heritage can, and, in fact, should, provide an important contribution to the current discussion and criticism of the interference of politics and economy with science. The 100th anniversary of Weber's death should inspire this reflection.

Notes

1 Theorists who pursue this line of reasoning include, among others, Hekman (1983); Turner and Factor (1984); Proctor (1991); Eliaeson (2002); Lassman and Velody (2015); Bruun (2016); Turner (2019).

2 The literature devoted to Weber's theory has undoubtedly been enriched by various analyses on the dissonance between the methodological and practical levels of values. The boundaries of this discussion are quite clear. On the one hand, it is obvious that having adopted Rickert's distinction, Weber consistently distinguishes between the methodological level of "relevance to values" and practical evaluation. At the same time, it is equally obvious that when describing the scientific investigation through the concept of vocation, Weber does not deprive it of evaluative foundations. While some commentators (Aron 1971; Mommsen 1974; Schluchter 2000; Beetham 1974) focused on isolating and highlighting the difficulties in reconciling the two dimensions of evaluation, others tried to systematize them (Bruun 2016). The separation approach, however orderly and clearly argued, is in my opinion quite problematic, because it makes it difficult to appreciate Weber's contribution to the understanding of methodology and science itself.

3 Weber's proposition rejects a strong assumption about an objective analysis of cultural life (scientific concept of objectivity) – in his view such analysis simply does not exist. Instead, he adopts its narrow meaning by seeking to establish the criteria for the selection, analysis, and an orderly description of the subject of scientific investigation (Weber 1903–1906/1988, 53; 1904/1949, 72).

4 Weber himself was aware of these weaknesses. He expressly considered the limited validity of methodological instruments (including the ideal types), for their objective validity is necessarily relative and problematic whenever the methodological instruments are regarded as a mere presentation of empirical data. However, if they are not characterized in this way, which Weber in fact proposes, they have great heuristic value. This is when they can be adopted to measure and compare empirical content (Weber 1904/1949, 92–93).

5 Beyond what is absolutely necessary, I would prefer to avoid discussing the relationship between Weber and Rickert. As is well known, the extent of the influence of Rickert's philosophy on Weber's concept of science is the subject of an ongoing debate. On the one hand, it has been suggested that Rickert had a direct and multi-faceted influence on Weberian concepts (including the concept of division of sciences, the distinction between the process of relevance to values and evaluation, etc.). Burger (1976) represents this view most strongly, while Nusser (1986) and Merz (1992) in a more moderate form. By contrast, others claim that Rickert's influence decreases as Weber's doctrine develops. While it is still noticeable in his earlier works, in Weber's later works we witness its adaptive modification and even a dramatic break. See Bruun (2001), Ringer (1997). For the purposes of this chapter, it is assumed that, although Weber adopts and consistently applies Rickert's distinction between the methodical sense of "reference to values" and the ideological evaluation, he definitely breaks with Rickert's philosophy as regards the concept of value itself (absolute values versus culturally conditioned values) and the very vision of science (universality of science versus particularism of science).

6 However, the "points of view" in investigating the culture are not voluntary. According to Weber, they are always subordinated to the two types of "adequacy"

(Turner 2019, 584–592). The first involves the adequacy from the point of view of causation, which is characteristic of natural sciences. The second type, which is more important from the point of view of social science, consists in the adequacy from the point of view of sense, where the investigator, by means of the understanding explanation (interpretation), seeks to establish a meaningful basis for a given behavior (in order to identify its evaluative motive). The lack of adequacy from the point of view of sense, even in the case of the highest regularity of a given course of events, means that we are dealing only with a statistical probability that cannot be fully understood and therefore scientifically explained (Weber 1922/1978, 11). Through the concept of adequacy from the point of view of sense, Weber introduces a kind of evaluative structure: in the cognitive process, the investigator relates the phenomenon being studied to the universal "cultural values" and exposes those relationships that are significant (important) to us.

7 Tenbruck's proposition (1959, 582–583) – in my opinion – is overly radical. According to his account, the entire methodological reflection is not the objective but a means of description.

8 The linking of reality with the ideas of values that give the reality its meaning, exposing and ordering these parts in terms of their cultural significance, involves pointing to the fundamental difference between (a) the analysis of reality, its laws, and ordering it according to the general concepts of natural sciences; and (b) the way of doing the humanities and developing ideal types (Hekman 1983, 14; Rossi 1987, 37–44).

9 Merz (1992, 230) adopts the term of epistemological subjectivism (*erkenntnistheoretischen Subjektivismus*).

References

Aron, R. 2019. *Main Currents in Sociological Thought*, vol. 2. London: Routledge.

Beetham, D. 1974. *Max Weber and the Theory of Modern Politics*. London: George Allen & Unwin.

Bruun, H.H. 2001. "Weber on Rickert: From Value Relation to Ideal Type." *Max Weber Studies* 1 (2): 139–160.

Bruun, H.H. 2016. *Science, Values and Politics in Max Weber's Methodology*. London: Routledge.

Bruun, H.H., and S. Whimster. 2012. "Introduction." In *Max Weber: Collected Methodological Writings*, edited by H.H. Bruun and S. Whimster, pp. xi–xxviii. London: Routledge.

Burger, T. 1976. *Max Weber's Theory of Concept Formation: History, Laws, and Ideal Types*. Durham, NC: Duke University Press.

Eliaeson, S. 2002. *Max Weber's Methodologies: Interpretation and Critique*. Cambridge: Polity.

Gostmann, P. 2015. "Intellectual Freedom: On the Political 'Gestalt' of Weber and Kelsen or Strauss's Critique of Social Science Revisited". In *The Foundation of the Juridico-Political: Concept Formation in Hans Kelsen and Max Weber*, edited by I. Bryan, P. Langford, and J. McGarry, pp. 91–114. London: Routledge.

Habermas, J. 1982. *Theory of Communicative Action*, vol. 1. Boston: Beacon Press.

Hekman, S.J. 1983. *Weber, the Ideal Type, and Contemporary Social Theory*. Notre Dame, IN: University of Notre Dame Press.

Jacobsen, B. 1999. *Max Weber und Friedrich Albert Lange*. Wiesbaden: Springer Fachmedien.

Lassman, P., and I. Velody. 2015. "Introduction." In *Max Weber's 'Science as a Vocation'*, edited by P. Lassman and I. Velody, pp. xiii–xvii. London: Routledge.

Merz, P-U. 1992. *Max Weber und Heinrich Rickert. Die erkenntniskritischen Grundlagen der verstehenden Soziologie.* Würzburg: Königshausen & Neumann.

Mommsen, W.J. 1974. *Max Weber und die deutsche Politik 1890–1920.* Tübingen: Mohr.

Nusser, K-H. 1986. *Kausale Prozesse und sinnerfassende Vernunft. Max Webers philosophische Fundierung der Soziologie und der Kulturwissenschaften.* Freiburg/München: Verlag Karl Alber.

Proctor, R.N. 1991. *Value-Free Science? Purity and Power in Modern Knowledge.* Cambridge, MA: Harvard University Press.

Ringer, F. 1997. *Max Weber's Methodology. The Unification of the Cultural and Social Sciences.* Cambridge, MA and London: Harvard University Press.

Rossi, P. 1987. *Vom Historismus zur historischen Sozialwissenschaft.* Frankfurt am Main: Suhrkamp.

Schluchter, W. 2000. *Individualismus, Verantwortungsethik und Vielfalt.* Göttingen: Velbrück Wissenschaft.

Shapin, S. 2019. "Weber's Science as a Vocation: A Moment in the History of 'Is' and 'Ought'." *Journal of Classical Sociology* 3: 290–307.

Tenbruck, F.H. 1959. "Die Genesis der Methodologie Max Webers." *Kölner Zeitschrift für Soziologie und Sozialpsychologie* 11: 573–630.

Turner, C. 1992. *Modernity and Politics in the Work of Max Weber.* London: Routledge.

Turner, S.P. 2019. "Causation, Value Judgement, Verstehen." In *The Oxford Handbook of Max Weber*, edited by E. Hanke *et al.*, pp. 575–592. Oxford: Oxford University Press.

Turner, S.P., and R.A. Factor. 1984. *Max Weber and the Dispute over Reason and Value.* London: Routledge & Kegan Paul.

Weber, M. 1903–1906/1988. "Roscher und Knies und die logischen Probleme der historischen Nationalökonomie." In *Max Weber: Gesammelte Aufsätze zur Wissenschaftslehre*, pp. 1–145. Tübingen: J.C.B. Mohr (Paul Siebeck).

Weber, M. 1904/1949. "'Objectivity' in Social Science and Social Policy." In *Max Weber on the Methodology of the Social Sciences*, edited by E.A. Shils and H.A. Finch, pp. 49–112. Illinois: Free Press of Glencoe.

Weber, M. 1906. "Zur Lage der burgerlischen Demokratie in Russland." *Archiv für Socialwissenschaft und Socialpolitik* 22: 234–353.

Weber, M. 1917/1949. "The Meaning of 'Ethical Neutrality' in Sociology and Economics." In *From Max Weber: Essays in Sociology*, edited by H.H. Gerth and C.W. Mills, pp. 1–47. New York: Oxford University Press.

Weber, M. 1918–1919/1946. "Science as a Vocation." In *From Max Weber: Essays in Sociology*, edited by H.H. Gerth and C.W. Mills, pp. 129–156. New York: Oxford University Press.

Weber, M. 1919/1946. "Politics as a Vocation." In *From Max Weber: Essays in Sociology*, edited by H.H. Gerth and C.W. Mills, pp. 77–128. New York: Oxford University Press.

Weber, M. 1922/1978. *Economy and Society*, edited by R.G.C. Wittich. Berkeley: University of California Press.

Weber, M. 1930/2005. *The Protestant Ethic and the Spirit of Capitalism.* London: Routledge.

Contributors

Heather Douglas is an associate professor in the Department of Philosophy at Michigan State University. She is a fellow of the Institute for Science, Society, and Policy at the University of Ottawa, member of the Virtual Institute for Responsible Innovation, and fellow of the American Association for the Advancement of Science. Her research interests are the relationship between science and democracy, the role of social and ethical values of science, science policy, communication of science, as well as objectivity in science. She has authored several articles as well as a monograph (*Science, Policy, and the Value-Free Ideal*, University of Pittsburgh Press, 2009). She is the editor of the University of Pittsburgh Press's new book series, *Science, Values, and the Public*.

Lidia Godek is an assistant professor in the Faculty of Philosophy at Adam Mickiewicz University, Poznań. She earned Bednarowski Trust Fellowships at the University of Aberdeen in 2018 and Fellowships at the Autonomous University of Barcelona in 2020. Her research focuses on Max Weber's philosophy and sociology of science. She is the author of several articles on Weber including "On deformational modelling. Max Weber's concept of idealization" (in *Idealization XIV: Models in Science*, Brill, 2016) and a monograph (*Max Weber's Philosophy of Stateness (Filozofia państwowości Maxa Webera)*, Institute of Philosophy AMU Press, 2013).

Péter Hartl is a research fellow at the Institute of Philosophy, Research Centre for the Humanities, Budapest, MTA BTK Lendület Morals and Science Research Group. His research focuses on epistemology and the history of philosophy (Hume, Michael Polanyi). He has published papers on Polanyi, Hume, and modal epistemology. He co-edited "The Value of Truth" special issue for *Synthese*. His monograph on Hume is under contract with Palgrave Macmillan.

Janet Kourany is an associate professor in the Department of Philosophy, University of Notre Dame. Her research areas include philosophy of science, science and social values, philosophy of feminism, and the new interdisciplinary area of ignorance studies. Her books include *Science and* *the Production of Ignorance: When the Quest for Knowledge is Thwarted* (co-edited with Martin Carrier; MIT Press, 2020); *Philosophy of Science after Feminism* (Oxford University Press, 2010); *The Challenge of the Social and the Pressure of Practice: Science and Values Revisited* (co-edited with Martin Carrier and Don Howard; University of Pittsburgh Press, 2008); *The Gender of Science* (Prentice Hall, 2002); *Feminist Philosophies* (co-edited with James Sterba and Rosemarie Tong; Prentice Hall, 1999, 1992); *Philosophy in a Feminist Voice* (Princeton University Press, 1998); and *Scientific Knowledge* (Wadsworth, 1998, 1987).

Hugh Lacey is a professor emeritus at Swarthmore College, and a member of the Working Group of Philosophy, History and Sociology of Science and Technology of the Institute of Advanced Studies at the Universidade de São Paulo. He held numerous visiting professorships including at the University of Melbourne; Pontificia Universiade Catolica, São Paulo; Rosemont College; Villanova University; Temple University; and the University of Pennsylvania. His book *Is Science Value-Free? Value and Scientific Understanding* (Routledge, 1999) is one of the classical works on this subject.

Phil Mullins is a professor emeritus who taught in an interdisciplinary humanities program at Missouri Western State University. For more than 20 years, he edited *Tradition and Discovery: The Polanyi Society Journal*, and currently is President of the Polanyi Society Board of Directors. He has published many essays on aspects of Polanyi's liberal political philosophy, his epistemology and philosophy of science.

Dustin Olson is a philosophy instructor at Luther College, University of Regina. His research focuses on social-political epistemology, belief control, philosophy of science, and the history of twentieth century analytic philosophy. He has published papers on metaphilosophy, epistemic agency, expert disagreement, the relationship between philosophy and science, and Bertrand Russell's philosophy.

Hans Radder is professor emeritus in the Department of Philosophy, VU University Amsterdam. His research focuses on the philosophy of science and technology, including their social, political, and ethical aspects. He is the author of *From Commodification to the Common Good: Reconstructing Science, Technology, and Society* (University of Pittsburgh Press, 2019), *The World Observed/The World Conceived* (University of Pittsburgh Press, 2006), and *In and About the World: Philosophical Studies of Science and Technology* (State University of New York Press, 1996). He also edited *The Commodification of Academic Research: Science and the Modern University* (University of Pittsburgh Press, 2010) and *The Philosophy of Scientific Experimentation* (University of Pittsburgh Press, 2003).

Jeroen Van Bouwel is a visiting professor at Ghent University. He has been a member of the Ghent Centre for Logic and Philosophy and Philosophy of Science for the past 20 years. His research focuses on social epistemology, philosophy of the social sciences, and the relations between science and democracy. He has published over 50 papers in refereed journals, edited a volume *The Social Sciences and Democracy* (Palgrave Macmillan, 2009), and co-authored a book *Scientific Explanation* (with Erik Weber and Leen De Vreese, Springer, 2013).

Index

Printed in the United States
by Baker & Taylor Publisher Services